Currents and Mesons

Chicago Lectures in Physics

currents and mesons

J. J. Sakurai

Based on notes compiled by George W. Barry

The University of Chicago Press
Chicago and London

Chicago Lectures in Physics Series

Experimental Superfluidity, by Russell J. Donnelly (1967)
Group Theory and Its Physical Applications, by Leopoldo M. Falicov (1966)
Elementary Particles, by Riccardo Levi Setti (1963)
Covalent Bonding in Crystals, Molecules, and Polymers,
by James C. Phillips (1969)
Currents and Mesons, by J. J. Sakurai (1969)

International Standard Book Number: 0–226–73383–1
Library of Congress Catalog Card Number: 69–15230
The University of Chicago Press, Chicago 60637
The University of Chicago Press, Ltd., London

Published 1969
Second Impression 1973
Printed in the United States of America

PREFACE

In the summer quarter of 1967 I gave a course on "Currents and Mesons" for advanced graduate students majoring in theoretical high-energy physics at the University of Chicago. A similar and somewhat shorter series of lectures was given earlier at Scuola Normale Superiore (Pisa) in the spring of 1967. Part of the course material (mainly concerned with vector mesons) was presented also in a series of lectures I gave at the Stanford Linear Accelerator Center in the winter of 1968.

The present set of notes, based on the Chicago version of my lectures, was prepared by a student of mine, George W. Barry. Little attempt has been made to put the notes in polished "book form".

I would like to thank Professors L. A. Radicati and G. Bernardini for their hospitality at Scuola Normale Superiore where much of the lecture material was prepared and George W. Barry for his tireless effort without which this project would not have been materialized.

J. J. S.

TABLE OF CONTENTS

III. Vector Meson Universality

History

Gauge principle and universality

Connection between vector meson dominance and universality.

Current-field identity (CFI) for ρ meson

Comparison with dispersion-theoretic treatment

Renormalized vs. unrenormalized ρ field

Determination of the ρ meson coupling constant

Gauge invariance and the $e m_\rho^2 / f_\rho$ presecription

$\omega - \varphi$ complex

Gauge field algebra

IV. PCAC and the Goldberger-Treiman Relation

Matrix elements for $\pi^\pm$ and neutron decay

Troubles with CAC (conserved axial-vector current)

Nambu's derivation of the G-T relation

Models satisfying PCAC

PCAC and pion pole dominance

PCAC and SU(3); generalized G-T relations

V. Soft Pion Processes

Review of the reduction technique

Single soft-pion emission (absorption) in strong-interaction processes

Adler consistency condition

CHAPTER I. REVIEW OF SU(3).

Quarks. The group SU(3) [or U(3)] was first introduced to particle physics within the framework of the symmetric Sakata model in 1960 (Ikeda, Ogawa, Ohnuki; Wess; Yamaguchi). In this model the physical p, n and Λ are taken as fundamental components from which all hadrons are to be constructed. Although the Sakata model is good for classifying the mesons, it fails for the baryons. Indeed, for the $1/2^+$ baryons the asymmetry is obvious because we have selected p, n and Λ as fundamental rather than some other triplet - say Λ, Ξ^0 and Ξ^-. What is worse, an attempt to classify the $3/2^+$ states was not so successful. Gell-Mann ('61) and Ne'eman ('61) abandoned this model in favor of the "eightfold way", which has led, among other things, to the successful predictions of Ω^-.

In the eightfold way none of the hadrons observed up to now belong to unitary triplets. However, it is profitable, for pedagogical purposes, to introduce a fermion triplet, called "quarks" (Gell-Mann '64; Zweig '64) because they have simpler transformation properties under SU(3) than the observed mesons and baryons. (The literal existence of the quarks is irrelevant for what we do in this course.) The quark triplet

$$q = \begin{pmatrix} p' \\ n' \\ \Lambda' \end{pmatrix}$$

transforms under SU(3) by the infinitesimal unitary transformation,

$$\left(1 + i \frac{\lambda_\alpha}{2} \varepsilon_\alpha\right) q \qquad\qquad \alpha = 1, \cdots, 8 ,$$

where

$$\left(1 + i \frac{\lambda_\alpha}{2} \varepsilon_\alpha\right)\left(1 - i \frac{\lambda_\alpha^+}{2} \varepsilon_\alpha\right) = 1 + O(\varepsilon_\alpha^2) .$$

The λ's are 3x3 traceless hermitian matrices and the ε's are infinitesimal real parameters. This is in direct analogy with the SU(2) isospin transformation,

$$(1 + i \frac{\tau_\alpha}{2} \varepsilon_\alpha) \begin{pmatrix} p \\ n \end{pmatrix},$$

where τ_3 distinguishes p from n, and τ_1 and τ_2 mix p and n. Their SU(3) analogs, the λ_α's, either mix or distinguish the p', n', and Λ'.

i) λ_1, λ_2, λ_3 mix or distinguish the p' and n' and are therefore expressible in terms of the τ_α's of SU(2).

$$\lambda_\alpha = \begin{pmatrix} \tau_\alpha & \begin{matrix} 0 \\ 0 \end{matrix} \\ \begin{matrix} 0 & 0 \end{matrix} & 0 \end{pmatrix} \qquad \alpha = 1,2,3$$

ii) λ_4, λ_5 mix p' and Λ'.

$$\lambda_4 = \begin{pmatrix} 0 & 0 & 1 \\ 0 & 0 & 0 \\ 1 & 0 & 0 \end{pmatrix} \qquad \lambda_5 = \begin{pmatrix} 0 & 0 & -i \\ 0 & 0 & 0 \\ i & 0 & 0 \end{pmatrix}$$

iii) λ_6, λ_7 mix n' and Λ'.

$$\lambda_6 = \begin{pmatrix} 0 & 0 & 0 \\ 0 & 0 & 1 \\ 0 & 1 & 0 \end{pmatrix} \qquad \lambda_7 = \begin{pmatrix} 0 & 0 & 0 \\ 0 & 0 & -i \\ 0 & i & 0 \end{pmatrix}$$

iv) λ_8 distinguishes Λ' from (p', n').

$$\lambda_8 = \frac{1}{\sqrt{3}} \begin{pmatrix} 1 & 0 & 0 \\ 0 & 1 & 0 \\ 0 & 0 & -2 \end{pmatrix}$$

The reader is encouraged to memorize the following useful diagram showing what the λ_α's do.

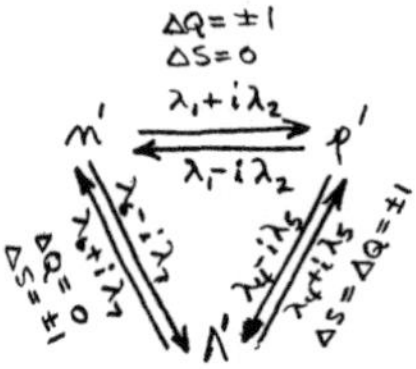

Properties of the λ Matrices

The λ matrices as well as being hermitian and traceless

$$\lambda_\alpha^\dagger = \lambda_\alpha$$
$$\mathrm{Tr}(\lambda_\alpha) = 0$$

are normalized by

$$\mathrm{Tr}(\lambda_\alpha \lambda_\beta) = 2\delta_{\alpha\beta} .$$

From the explicit expressions for the λ_α's we can compute the structure constants $f_{\alpha\beta\gamma}$ of the SU(3) commutation relations,

$$[\lambda_\alpha , \lambda_\beta] = 2i f_{\alpha\beta\gamma} \lambda_\gamma$$

Why twice $f_{\alpha\beta\gamma}$? Well, λ_1, λ_2, and λ_3 behave like the τ_α's which in turn behave like angular momentum.

$$\left[\frac{\tau_\alpha}{2} , \frac{\tau_\beta}{2}\right] = i\,\varepsilon_{\alpha\beta\gamma} \frac{\tau_\gamma}{2}$$

So we need the 2 for $f_{\alpha\beta\gamma}$ to agree with $\varepsilon_{\alpha\beta\gamma}$. This shows that a more convenient form is

$$\left[\frac{\lambda_\alpha}{2} , \frac{\lambda_\beta}{2}\right] = i f_{\alpha\beta\gamma} \frac{\lambda_\gamma}{2} .$$

From the anticommutator we compute the $d_{\alpha\beta\gamma}$'s which are defined by

$$\{\lambda_\alpha , \lambda_\beta\} = \frac{4}{3}\delta_{\alpha\beta} + 2 d_{\alpha\beta\gamma} \lambda_\gamma \qquad \text{(no i).}$$

An alternative definition is

$$\{\lambda_\alpha , \lambda_\beta\} = 2 d_{\alpha\beta\gamma} \lambda_\gamma \qquad \alpha, \beta, \gamma = 0, 1, \cdots, 8 ,$$

where

$$d_{\alpha\beta 0} = d_{\alpha 0 \beta} = d_{\beta\alpha 0} = d_{\beta 0 \alpha} = d_{0\alpha\beta} = d_{0\beta\alpha} = \sqrt{\tfrac{2}{3}}\,\delta_{\alpha\beta} , \quad \lambda_0 = \sqrt{\tfrac{2}{3}} .$$

The structure constants can simply be evaluated by

$$\mathrm{Tr}(\lambda_\gamma [\lambda_\alpha , \lambda_\beta]) = 4i f_{\alpha\beta\gamma}$$
$$\mathrm{Tr}(\lambda_\gamma \{\lambda_\alpha , \lambda_\beta\}) = 4 d_{\alpha\beta\gamma} .$$

We now list some properties of the d's and f's.

i) The d's and f's are real.

$$2if_{\alpha\beta\gamma}\lambda_\gamma = [\lambda_\alpha, \lambda_\beta] = -[\lambda_\alpha, \lambda_\beta]^+ = 2if^*_{\alpha\beta\gamma}\lambda^+_\gamma = -2if^*_{\alpha\beta\gamma}\lambda_\gamma$$

therefore $f^*_{\alpha\beta\gamma} = f_{\alpha\beta\gamma}$

A similar proof applies for the d's.

ii) The f's are totally antisymmetric; the d's are totally symmetric.

Proof: The antisymmetry of $f_{\alpha\beta\gamma}$ with respect to α, β is obvious from the definition. Antisymmetry in β, γ follows from

$$4if_{\alpha\beta\gamma} = \mathrm{Tr}(\lambda_\gamma[\lambda_\alpha, \lambda_\beta]) = -\mathrm{Tr}(\lambda_\beta[\lambda_\alpha, \lambda_\gamma]) = -4if_{\alpha\gamma\beta}.$$

A similar proof applies for the d's.

iii) The d's vanish whenever an odd number of 2's, 5's and 7's occur. (e.g. $d_{157} \neq 0$ but $d_{156} = 0$, $d_{257} = 0$)
The f's vanish whenever an even number of 2's, 5's and 7's occur. (e.g. $f_{257} \neq 0$ but $f_{267} = 0$)

Proof: λ_2, λ_5, λ_7 are purely imaginary - the others are real. Define ε_α so that

$$\lambda^*_\alpha = \lambda^T_\alpha = \varepsilon_\alpha\lambda_\alpha \qquad (\alpha \text{ not summed}).$$

Note

$$\varepsilon_\alpha = \begin{cases} +1 & \alpha = 1,3,4,6,8 \\ -1 & \alpha = 2,5,7. \end{cases}$$

Taking the complex conjugate of $[\lambda_\alpha, \lambda_\beta] = 2if_{\alpha\beta\gamma}$ we have

$$\varepsilon_\alpha\varepsilon_\beta[\lambda_\alpha, \lambda_\beta] = -2if_{\alpha\beta\gamma}(\varepsilon_\gamma\lambda_\gamma) \quad (\alpha, \beta \text{ not summed}),$$

$$\varepsilon_\alpha\varepsilon_\beta\varepsilon_\gamma f_{\alpha\beta\gamma} = -f_{\alpha\beta\gamma}.$$

iv) $(1/\sqrt{3})$ times integer appears only when one of the indices is an 8.

Proof: Obvious

Non-zero elements of the d's and the f's are tabulated in Appendix.

Electric Charge of Quarks

The Gell-Mann-Nishijima rule,

$$Q^{(el)} = T_3 + \frac{Y}{2},$$

relates the electric charge to the third component of the i-spin (T_3) (the eigenvalue of $\lambda_3/2$) and the hypercharge (Y). If we require Y to be proportional to the eigenvalue of λ_8 (so that electric charge transforms like a member of an octet, i.e. no singlet component), then $Q^{(el)}$ can only be an eigenvalue of $\frac{\lambda_3}{2} + \frac{1}{\sqrt{3}}\frac{\lambda_8}{2}$ (Note that the Y's of the isosinglet quark and the isodoublet quark must differ by 1.) More precisely,

$$Q^{(el)} = \int d^3x\, q^\dagger \left(\frac{\lambda_3}{2} + \frac{1}{\sqrt{3}}\frac{\lambda_8}{2}\right) q = Q^3 + \frac{1}{\sqrt{3}} Q^8,$$

$$Q^\alpha \equiv \int d^3x\, q^\dagger(x) \frac{\lambda_\alpha}{2} q(x),$$

$$\frac{\lambda_3}{2} + \frac{1}{\sqrt{3}}\frac{\lambda_8}{2} = \begin{pmatrix} 2/3 & 0 & 0 \\ 0 & -1/3 & 0 \\ 0 & 0 & -1/3 \end{pmatrix}.$$

The fractional electric charge for the triplet is an unavoidable consequence of requiring the Y defined by the Gell-Mann-Nishijima relation to be proportional to λ_8 (or that $Q^{(el)}$ has no singlet component). If you don't like fractional charge, then you can:

i) Abandon the idea that $Q^{(el)}$ transforms like an octet - or

ii) Disbelieve in triplets.

SU(2) Subgroups

(Meshkov, Levinson and Lipkin '63)

The following SU(2) subgroups of SU(3) satisfy the angular momentum commutation relations

$$\left[\frac{\lambda'_\alpha}{2}, \frac{\lambda'_\beta}{2}\right] = i\,\varepsilon_{\alpha\beta\gamma}\frac{\lambda'_\gamma}{2} \qquad (\alpha, \beta, \gamma = 1, 2, 3).$$

T spin: $\left[\frac{\lambda_1}{2}, \frac{\lambda_2}{2}, \frac{\lambda_3}{2}\right]$

U spin: $\left[\frac{\lambda_6}{2}, \frac{\lambda_7}{2}, \frac{1}{2}\left(-\frac{\lambda_3}{2} + \sqrt{3}\,\frac{\lambda_8}{2}\right)\right]$

V spin: $\left[\frac{\lambda_4}{2}, \frac{\lambda_5}{2}, \frac{1}{2}\left(\frac{\lambda_3}{2} + \sqrt{3}\,\frac{\lambda_8}{2}\right)\right]$

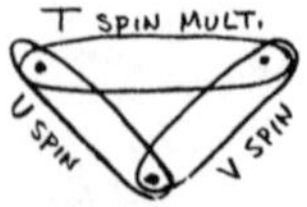

i) $SU(2)_T$ operators commute with $Y = \frac{2}{\sqrt{3}} Q^8$

ii) $SU(2)_U$ " " " $Q^{el} = Q^3 + \frac{1}{\sqrt{3}} Q^8$

iii) $SU(2)_V$ " " " $Q^3 - \frac{1}{\sqrt{3}} Q^8$

A consequence of (ii) is that members of a given U spin multiplet have the same electromagnetic properties, eg. $\mu(\Sigma^+) = \mu(p)$.

Another possible subgroup is obtained by commuting linear combinations of the charge-changing components of the T spin and V spin,

$$\left[\frac{\lambda'_1}{2}, \frac{\lambda'_2}{2}, \frac{\lambda'_3}{2}\right] = \left[\frac{\lambda_1}{2}\cos\theta + \frac{\lambda_4}{2}\sin\theta, \frac{\lambda_2}{2}\cos\theta + \frac{\lambda_5}{2}\sin\theta, \right.$$
$$\left. \frac{\lambda_3}{2}\cos^2\theta + \frac{1}{2}\left(\frac{\lambda_3}{2} + \sqrt{3}\,\frac{\lambda_8}{2}\right)\sin^2\theta - \frac{1}{2}\lambda_6 \sin\theta\cos\theta\right].$$

In semileptonic weak interactions the leptonic current (which is charge-changing) is coupled to the hadronic charge-changing current, which in its most general form is a linear combination

of the $1 \pm i2$ components of the T spin and V spin. So in general the hadronic current transforms as $\bar{q}\left\{\begin{matrix}\gamma_\mu \\ \gamma_\mu\gamma_5\end{matrix}\right\}\left(\frac{\lambda_1'}{2} \pm i\frac{\lambda_2'}{2}\right)q$.
Using this idea, we may formulate Cabibbo's universality.

Mesons

We now express the mesons (octet and singlet) in terms of the 12 component quark field (a Dirac spinor for each of the triplet p', n', Λ'). The mesons visualized as a bound system of a quark and antiquark correspond to the bilinear form $\bar{q}\,\mathcal{O}\,q$ where the 12x12 matrix $\mathcal{O}$ is the direct product of the 4x4 Dirac matrices and the 3x3 unitary spin matrices ($\mathcal{O} = \Gamma \otimes \lambda$).
The pseudoscalar mesons have the same transformation properties as

$\bar{q}\frac{i\gamma_5}{\sqrt{3}}q$: singlet $\eta(960)$

$\bar{q}\frac{i\gamma_5\lambda_\alpha}{\sqrt{2}}q$: Octet $\pi, K, \bar{K}, \eta(549)$

In short the meson fields transform as

$\varphi^0 \sim \frac{\bar{q}q}{\sqrt{3}}$: Singlet

$\varphi^\alpha \sim \frac{\bar{q}\lambda_\alpha q}{\sqrt{2}}$: Octet

(The $\sim$ means "transforms as", and we omit the $i\gamma_5$ for simplicity.)
The φ^α need not have a definite T_3 or $Q^{(el)}$ but certain linear combinations do. For example:

$$\pi^0 \sim \bar{q}\frac{\lambda_3}{\sqrt{2}}q = \frac{\bar{p}'p' - \bar{n}'n'}{\sqrt{2}}$$

$$\pi^\pm \sim \bar{q}\left(\frac{\lambda_1 \mp i\lambda_2}{2}\right)q = \left\{\begin{matrix}\bar{n}'p' \\ \bar{p}'n'\end{matrix}\right\}$$

$$K^\pm \sim \bar{q}\left(\frac{\lambda_4 \mp i\lambda_5}{2}\right)q = \left\{\begin{matrix}\bar{\Lambda}'p' \\ \bar{p}'\Lambda'\end{matrix}\right\}$$

$$\left\{\frac{K^0}{\bar{K}^0}\right\} \sim \bar{q}\frac{(\lambda_6 \mp i\lambda_7)}{2} q = \left\{\frac{\bar{\Lambda}' n'}{\bar{n}' \Lambda'}\right\}$$

$$\eta(549) \sim \bar{q}\frac{\lambda_8}{\sqrt{2}} q = \frac{\bar{p}'p' + \bar{n}'n' - 2\bar{\Lambda}'\Lambda'}{\sqrt{6}}$$

$$\eta(960) \sim \frac{\bar{q}q}{\sqrt{3}} = \frac{\bar{p}'p' + \bar{n}'n' + \bar{\Lambda}'\Lambda'}{\sqrt{3}} .$$

The field operator π^+ annihilates π^+, or creates π^-. This is to be compared with $\bar{n}'p$ which annihilates p' and $\bar{n}'$ (or annihilates p' and creates n' etc.). The sum of the charges of the constituents

$$Q\ (p') + Q\ (\bar{n}') = \frac{2}{3} + (-1)\ (-\tfrac{1}{3}) = 1 = Q\ (\pi^+)$$

gives the desired charge of the π^+, as it must.

Vector vs Tensor Notation

Under infinitesimal SU(3) transformation we have

$$q \longrightarrow (1 + i\,\tfrac{\lambda_\beta}{2}\varepsilon_\beta)\,q$$

$$\bar{q} \longrightarrow \bar{q}(1 - i\,\tfrac{\lambda_\beta}{2}\varepsilon_\beta)$$

$$\bar{q}\lambda_\alpha q \longrightarrow \bar{q}\lambda_\alpha q + i\frac{\varepsilon_\beta}{2}\bar{q}[\lambda_\alpha, \lambda_\beta]q + O(\varepsilon_\beta^2)$$

$$= \bar{q}\lambda_\alpha q - f_{\alpha\beta\gamma}\varepsilon_\beta\bar{q}\lambda_\gamma q$$

$$\varphi^\alpha \longrightarrow \varphi^\alpha - f_{\alpha\beta\gamma}\varepsilon_\beta\varphi^\gamma .$$

Thus the transformation property of φ^α is analogous to the "vector" in 3 dimensions

$$\vec{V} \longrightarrow \vec{V} - (\delta\vec{\omega}) \times \vec{V} \quad , \quad [\ V_i \to V_i - \varepsilon_{ijk}(\delta\vec{\omega})_j V_k]$$

and the $f_{\alpha\beta\gamma}$ acts on octets as ε_{ijk} does on vectors.

The octet can also be represented as a traceless 3x3 tensor (matrix). From the triplet (q_a) and antitriplet ($\bar{q}^a$) (a = 1,2,3) we can form the tensor product $\bar{q}^b q_a$.

This product can be decomposed into its trace and a traceless 8 component object.

$$\bar{q}^a q_a \qquad : \text{singlet}$$

$$\bar{q}^b q_a - \tfrac{1}{3}\delta_a^b \bar{q}^c q_c \quad : \text{octet}$$

What is the connection between the two notations? Writing the octet tensor as a matrix, we have

$$\mathcal{M}_{ab} = \bar{q}^b q_a - \tfrac{1}{3}\delta_a^b \bar{q}^c q_c \qquad (\text{e.g. } \mathcal{M}_{23} = \bar{q}^3 q_2 = \bar{\lambda}' n' = K^0),$$

$$\mathcal{M} = \begin{pmatrix} \frac{\pi^0}{\sqrt{2}} + \frac{\eta}{\sqrt{6}} & \pi^+ & K^+ \\ \pi^- & -\frac{\pi^0}{\sqrt{2}} + \frac{\eta}{\sqrt{6}} & K^0 \\ K^- & \bar{K}^0 & -\frac{2\eta}{\sqrt{6}} \end{pmatrix} = \frac{1}{\sqrt{2}} \lambda_\alpha \varphi_\alpha .$$

Thus the 3x3 λ matrices rearrange the various components of the octet "vector" into an octet "tensor".

Charge Conjugation Properties

There is no phase ambiguity in defining the charge conjugation properties of the π^0 and η since both of these can decay into two gammas.

$$\pi^0 \overset{C}{\longleftrightarrow} \pi^0$$

$$\eta \overset{C}{\longleftrightarrow} \eta$$

For the others a convenient choice of phase is +1 (just convention - no experiment requires this).

$$\pi^+ \overset{C}{\longleftrightarrow} \pi^-$$

$$K^+ \overset{C}{\longleftrightarrow} K^-$$

$$K^0 \overset{C}{\longleftrightarrow} \bar{K}^0$$

In short
$$\eta \xrightarrow{C} \eta^T .$$

With this convention φ^α that corresponds to $\bar{q} i \gamma_5 \lambda_\alpha q$, however, cannot always have the phase factor +1.

$$\bar{q} i \gamma_5 \lambda_\alpha q \xrightarrow{C} \begin{cases} \bar{q} i \gamma_5 \lambda_\alpha q & \alpha = 1,3,4,6,8 \\ -\bar{q} i \gamma_5 \lambda_\alpha q & \alpha = 2,5,7 \end{cases}$$

This is no surprise if we recall that the λ_α's are imaginary for α = 2,5,7 and real for α = 1,3,4,6,8; after all charge conjugation is just complex conjugation (apart from a shuffling of the Dirac spinor components, which does not occur in the Majorana representation). In general, given any octet with B=0, it is possible to define $\mathcal{C}$ which characterizes the multiplet. It is simply equal to the C (ordinary charge conjugation parity) of the 3rd and 8th components (e.g. π^0 and η).

$$\mathcal{C} = \begin{cases} +1 \text{ for pseudoscalar octet} \\ -1 \text{ for vector octet} \end{cases}$$

With the particular choice of phase indicated above, we obtain

$$\mathcal{C} = \begin{cases} +C, \text{ for } 1,3,4,6,8 \\ -C, \text{ for } 2,5,7 \end{cases}$$

As an application of these concepts, we examine the behavior of K^0 and $\bar{K}^0$ under CP.

$$K^0 \sim \bar{q} i \gamma_5 \frac{(\lambda_6 - i\lambda_7)}{\sqrt{2}} q \xrightarrow{CP} -\bar{q} i \gamma_5 \frac{(\lambda_6 + i\lambda_7)}{\sqrt{2}} q \sim -\bar{K}^0$$

(The minus sign comes from the parity operation on the pseudoscalar state). Hence the CP eigenstates of K^0, $\overline{K}^0$ are

$$CP=-1: \quad \frac{K^0+\overline{K}^0}{\sqrt{2}} = \varphi^6 \equiv K_2$$

$$CP=+1: \quad -\frac{K^0-\overline{K}^0}{i\sqrt{2}} = \varphi^7 \equiv K_1 .$$

Our phase convention should be contrasted with the "usual" phase convention based on $CP|K^0\rangle = |\overline{K}^0\rangle$, $CP|\overline{K}^0\rangle = |K^0\rangle$, which leads to

$$CP=+1: \quad \frac{K^0+\overline{K}^0}{\sqrt{2}} = K_1$$

$$CP=-1: \quad \frac{K^0-\overline{K}^0}{\sqrt{2}} = K_2 .$$

Baryon Octet

A baryon is made up of three quarks. The product of three quarks can be split up into irreducible representations:

$$3\text{x}3\text{x}3 = 1+8+8+10 .$$

This reduction is important for discussing SU(6) or the non-relativistic quark model, but as far as the transformation properties of the $J = \frac{1}{2}$ baryons under SU(3) are concerned, it is sufficient to remember that they behave like the mesons. So by making a correspondence with the pseudoscalar octet ($\pi \to \Sigma$, $\eta \to \Lambda$, etc.) we get the 3x3 traceless tensor

$$\mathcal{B} = \begin{pmatrix} \frac{\Sigma^0}{\sqrt{2}}+\frac{\Lambda^0}{\sqrt{6}} & \Sigma^+ & p \\ \Sigma^- & -\frac{\Sigma^0}{\sqrt{2}}+\frac{\Lambda^0}{\sqrt{6}} & n \\ \Xi^- & \Xi^0 & -\frac{2\Lambda^0}{\sqrt{6}} \end{pmatrix} = \frac{1}{\sqrt{2}} \lambda_\alpha \psi_\alpha .$$

Moreover,

$$\overline{\mathcal{B}} = \begin{pmatrix} \frac{\overline{\Sigma}^0}{\sqrt{2}}+\frac{\overline{\Lambda}^0}{\sqrt{6}} & \overline{\Sigma}^- & \overline{\Xi}^- \\ \overline{\Sigma}^+ & -\frac{\overline{\Sigma}^0}{\sqrt{2}}+\frac{\overline{\Lambda}^0}{\sqrt{6}} & \overline{\Xi}^0 \\ \overline{p} & \overline{n} & -\frac{2\overline{\Lambda}^0}{\sqrt{6}} \end{pmatrix} = \frac{1}{\sqrt{2}} \lambda_\alpha \overline{\psi}_\alpha .$$

Note that

$$\overline{\psi}_\alpha \neq \overline{(\psi_\alpha)} .$$

Construction of Invariant Couplings

In tensor notation we have, under infinitesimal SU(3) transformations,

$$\mathcal{M} = \frac{1}{\sqrt{2}}\lambda_\alpha \varphi_\alpha \longrightarrow \frac{1}{\sqrt{2}}(\lambda_\alpha \varphi^\alpha - \varepsilon_\beta \lambda_\alpha f_{\alpha\beta\gamma}\varphi^\gamma)$$
$$= \mathcal{M} + i\frac{\varepsilon_\beta}{2}[\lambda_\beta, \mathcal{M}] = U\mathcal{M}U^+,$$

where

$$U = 1 + i\varepsilon_\beta \frac{\lambda_\beta}{2}$$

and we have used

$$-\lambda_\alpha f_{\alpha\beta\gamma} = -f_{\beta\gamma\alpha}\lambda_\alpha = \frac{i}{2}[\lambda_\beta, \lambda_\gamma].$$

Similarly,

$$\mathcal{B} \longrightarrow U\mathcal{B}U^+.$$

There are two invariant baryon-meson couplings (SU(3) singlets),

$$\mathrm{Tr}\ (\bar{\mathcal{B}}\mathcal{M}\mathcal{B}) \text{ and } \mathrm{Tr}\ (\bar{\mathcal{B}}\mathcal{B}\mathcal{M}).$$

It is better, however, to consider certain linear combinations of these.

(i) F-type coupling: $\mathrm{Tr}\ (\bar{\mathcal{B}}\mathcal{B}\mathcal{M} - \bar{\mathcal{B}}\mathcal{M}\mathcal{B})$

$$= \frac{1}{2^{3/2}}\mathrm{Tr}\,(\lambda_\alpha[\lambda_\beta, \lambda_\gamma])\bar{\psi}_\alpha \psi_\beta \varphi_\gamma$$
$$= \sqrt{2}\, i f_{\alpha\beta\gamma}\bar{\psi}_\alpha \psi_\beta \varphi_\gamma .$$

This expression says that the meson octet is coupled with an antisymmetric octet formed from a bilinear combination of $\bar{\psi}_\alpha$ and ψ_β.

(ii) D-type coupling: $\mathrm{Tr}\ (\bar{\mathcal{B}}\mathcal{B}\mathcal{M} + \bar{\mathcal{B}}\mathcal{M}\mathcal{B}) =$

$$= \sqrt{2}\, d_{\alpha\beta\gamma}\bar{\psi}_\alpha \psi_\beta \varphi_\gamma .$$

Here the φ_γ interact with a symmetric octet made from $\bar{\psi}_\alpha$ and ψ_β.

CHAPTER II. DIVERGENCE CONDITION AND CURRENT COMMUTATION RELATIONS (CCR)

Currents and charges in the free-field quark model

In the following we use the free-field quark model as a guide in deriving the properties of hadronic currents. The free quark Lagrangian,

$$\mathcal{L} = -(\bar{q}\gamma_\mu \partial_\mu q + m\bar{q}q),$$

gives the equations of motion,

$$\gamma_\mu \partial_\mu q + mq = 0,$$
$$-\partial_\mu \bar{q}\gamma_\mu + m\bar{q} = 0.$$

(The metric used here is $a \cdot b = \vec{a}\cdot\vec{b} + a_4 b_4 = \vec{a}\cdot\vec{b} - a_0 b_0$. No $g_{\mu\nu}$ is necessary.) Multiplying the first equation on the left by $\bar{q}\lambda_\alpha$, the second on the right by $\lambda_\alpha q$, and subtracting we get (assuming the quark masses are degenerate; $m_{p'} = m_{n'} = m_{\Lambda'} = m$)

$$\partial_\mu [\bar{q}\gamma_\mu \lambda_\alpha q] = 0.$$

This defines a divergenceless current.

$$j_\mu^\alpha = i\bar{q}\gamma_\mu \frac{\lambda_\alpha}{2} q \quad , \quad \partial_\mu j_\mu^\alpha = 0$$

These conserved currents give rise to constants of the motion (one for each α). To see this we merely note

$$\int d^3x \, \partial_\mu j_\mu^\alpha = 0$$

also implies

$$\frac{d}{dt}\int d^3x \, j_0^\alpha = 0 \qquad\qquad (j_0^\alpha = -i j_4^\alpha)$$

since $\int d^3x\, \partial_k j_k^\alpha = 0$ by Gauss' theorem. Hence we have eight constants of the motion:

$$Q^\alpha \equiv \int d^3x\, j_0^\alpha = \int d^3x\, \bar{q}\gamma_4 \frac{\lambda_\alpha}{2} q = \int d^3x\, q^\dagger \frac{\lambda_\alpha}{2} q,$$

$$\frac{dQ^\alpha}{dt} = 0 .$$

By considering λ_0 which is proportional to the identity matrix we see that $\frac{1}{\sqrt{6}}\int d^3x\, \bar{q}\gamma_4 q = \int d^3x\, j_0^0(x)$ is also a constant of the motion.

To get the axial vector currents, repeat the same procedure with $\gamma_5\lambda_\alpha$ in place of λ_α and add.

$$\partial_\mu(\bar{q}\gamma_\mu\gamma_5\lambda_\alpha q) = 2m\bar{q}\gamma_5\lambda_\alpha q$$

$$\partial_\mu j_{5\mu}^\alpha = i\, m\, \bar{q}\gamma_5\lambda_\alpha q\,, \qquad j_{5\mu}^\alpha = i\bar{q}\gamma_\mu\gamma_5 \frac{\lambda_\alpha}{2} q$$

The axial vector current is not conserved unless m = 0, and its divergence is in general a pseudoscalar density.

If the quark mass is zero, we have 16 + 2 conserved currents.

$$j_\mu^\alpha = i\bar{q}\gamma_\mu \frac{\lambda_\alpha}{2} q \qquad\qquad j_{5\mu}^\alpha = iq\gamma_\mu\gamma_5 \frac{\lambda_\alpha}{2} q$$

$$j_\mu^0 = i\bar{q}\gamma_\mu \frac{\lambda_0}{2} q = \frac{i}{\sqrt{6}}\bar{q}\gamma_\mu q \qquad\qquad j_{5\mu}^0 = \frac{i}{\sqrt{6}}\bar{q}\gamma_\mu\gamma_5 q \qquad \left(\lambda_0 = \sqrt{\tfrac{2}{3}}\right)$$

$$\partial_\mu j_\mu^\alpha = \partial_\mu j_\mu^0 = \partial_\mu j_{5\mu}^\alpha = \partial_\mu j_{5\mu}^0 = 0$$

These local conservation laws imply corresponding global conservation laws; i.e. Q^{α}, Q_5^{α}, Q^0, Q_5^0 are all constants of the motion. If $m \neq 0$, then

$$\partial_\mu j_\mu^{\alpha,0} = 0 \quad \text{but} \quad \partial_\mu j_{5\mu}^{\alpha,0} \neq 0 .$$

and we have only 8 + 1 conserved currents.

In the following we loosely refer to $Q^{\alpha,0}$, $Q_5^{\alpha,0}$ as "charges" and $j_\mu^{\alpha,0}$, $j_{5\mu}^{\alpha,0}$ as "currents."

Gell-Mann-Lévy method

We now take a more systematic approach to conserved and partially conserved currents using the Gell-Mann-Lévy variational method. We consider the change in the free-quark Lagrangian, $\mathcal{L} = -\bar{q}(\gamma_\mu \partial_\mu + m)q$, due to the transformation

$$q_\tau \longrightarrow q_\tau + \varepsilon(x) F_\tau(q) \qquad (\tau = 1, \cdots, 12)$$

$$\partial_\mu q_\tau \longrightarrow \partial_\mu q_\tau + \partial_\mu \varepsilon \, F_\tau + \varepsilon \partial_\mu F_\tau ,$$

where $F_\tau(q)$ is a linear functional of the quark-field operator, and $\varepsilon(\vec{x})$ is an infinitesimal space-time function. We get

$$\frac{\delta \mathcal{L}}{\delta \varepsilon} = \frac{\delta \mathcal{L}}{\delta q_\tau}\frac{\delta q_\tau}{\delta \varepsilon} + \frac{\delta \mathcal{L}}{\delta(\partial_\mu q_\tau)}\frac{\delta(\partial_\mu q_\tau)}{\delta \varepsilon}$$

$$= \frac{\delta \mathcal{L}}{\delta q_\tau} F_\tau + \frac{\delta \mathcal{L}}{\delta(\partial_\mu q_\tau)} \partial_\mu F_\tau ,$$

$$\frac{\delta \mathcal{L}}{\delta(\partial_\mu \varepsilon)} = \frac{\delta \mathcal{L}}{\delta(\partial_\mu q_\tau)}\frac{\delta(\partial_\mu q_\tau)}{\delta(\partial_\mu \varepsilon)} = \frac{\delta \mathcal{L}}{\delta(\partial_\mu q_\tau)} F_\tau .$$

Differentiating the last equation and using the Euler-Lagrange equation

$$\partial_\mu\left(\frac{\delta\mathcal{L}}{\delta(\partial_\mu q_r)}\right) = \frac{\delta\mathcal{L}}{\delta q_r}$$

we obtain

$$\partial_\mu\left(\frac{\delta\mathcal{L}}{\delta(\partial_\mu\varepsilon)}\right) = \frac{\delta\mathcal{L}}{\delta\varepsilon} \qquad \text{(Gell-Mann-Lévy equation).}$$

If $\frac{\delta\mathcal{L}}{\delta\varepsilon} = 0$, then there is a conserved current, viz. $-\frac{\delta\mathcal{L}}{\delta(\partial_\mu\varepsilon)}$. The above equation looks like the Euler-Lagrange equation. But the similarity is a little misleading since $\varepsilon(x)$ is a gauge variable function and not a dynamical variable.

We now apply this formalism to certain quark transformations.

Example (1)

$$q \rightarrow (1 + i\,\varepsilon_\alpha \tfrac{\lambda_\alpha}{2})q$$

$$\bar{q} = q^\dagger\gamma_4 \rightarrow \bar{q}(1 - i\,\varepsilon_\alpha \tfrac{\lambda_\alpha}{2})$$

Here α is not summed because we are talking about a particular α, e.g. an isospin rotation about the 3rd axis. Thus

$$-\bar{q}\gamma_\mu\partial_\mu q \rightarrow -\bar{q}(1 - i\,\varepsilon_\alpha\tfrac{\lambda_\alpha}{2})\gamma_\mu\partial_\mu(1 + i\,\varepsilon_\alpha\tfrac{\lambda_\alpha}{2})q$$

$$= -\bar{q}\gamma_\mu\partial_\mu q - i\bar{q}\gamma_\mu\tfrac{\lambda_\alpha}{2}q\,\partial_\mu\varepsilon_\alpha$$

$$-\bar{q}q \rightarrow -\bar{q}(1 - i\varepsilon_\alpha\tfrac{\lambda_\alpha}{2})(1 + i\varepsilon_\alpha\tfrac{\lambda_\alpha}{2})q = -\bar{q}q$$

So for the free quark Lagrangian, $\mathcal{L} = -\bar{q}(\gamma_\mu\partial_\mu + m)q$,

$$\frac{\delta \mathcal{L}}{\delta \epsilon_\alpha} = 0 \quad \text{and} \quad \frac{\delta \mathcal{L}}{\delta(\partial_\mu \epsilon_\alpha)} = -i \bar{q} \gamma_\mu \frac{\lambda_\alpha}{2} q .$$

If we identify

$$j^\alpha_\mu = -\frac{\delta \mathcal{L}}{\delta(\partial_\mu \epsilon_\alpha)} = i \bar{q} \gamma_\mu \frac{\lambda_\alpha}{2} q ,$$

then j^α_μ is conserved ($\partial_\mu j^\alpha_\mu = 0$) since $\frac{\delta \mathcal{L}}{\delta \epsilon_\alpha} = 0$.

Example (2)

$$q \longrightarrow (1 + i \epsilon_\alpha \gamma_5 \frac{\lambda_\alpha}{2}) q$$

$$\bar{q} = q^\dagger \gamma_4 \longrightarrow \bar{q} (1 + i \epsilon_\alpha \gamma_5 \frac{\lambda_\alpha}{2})$$

(+ not - this time, since γ_5 anticommutes with γ_4)

$$-\bar{q} \gamma_\mu \partial_\mu q \longrightarrow -\bar{q} \gamma_\mu \partial_\mu q - i \partial_\mu \epsilon_\alpha \bar{q} \gamma_\mu \gamma_5 \frac{\lambda_\alpha}{2} q$$

$$-\bar{q} q \longrightarrow -\bar{q} q - i \epsilon_\alpha \bar{q} \gamma_5 \lambda_\alpha q$$

So

$$\frac{\delta \mathcal{L}}{\delta(\partial_\mu \epsilon_\alpha)} = -i \bar{q} \gamma_\mu \gamma_5 \frac{\lambda_\alpha}{2} q , \qquad \frac{\delta \mathcal{L}}{\delta \epsilon} = -i m \bar{q} \gamma_5 \lambda q ,$$

and we get axial-vector currents which are not divergenceless

$$j^\alpha_{5\mu} = -\frac{\delta \mathcal{L}}{\delta(\partial_\mu \epsilon_\alpha)} = i \bar{q} \gamma_\mu \gamma_5 \frac{\lambda_\alpha}{2} q ,$$

$$\partial_\mu j^\alpha_{5\mu} = i m \bar{q} \gamma_5 \lambda_\alpha q .$$

Symmetry Breaking Interaction and Divergence Conditions

Let us apply the transformations $q \to \left[1 + i\varepsilon_\alpha \left\{ \begin{matrix} 1 \\ \gamma_5 \end{matrix} \right\} \frac{\lambda_\alpha}{2}\right] q$ when interactions (or symmetry breaking mechanisms) are present. As long as the interaction Lagrangian does not involve derivatives, the currents generated are of the same form, since only the $-\bar{q}\gamma_\mu \partial_\mu q$ term contributes to the LHS of the G-M-Lévy equation. The RHS, however, is modified in general.

The introduction of a mass difference term into the Lagrangian (SU(3) breaking, medium-strong "interaction")

$$\delta m\, \bar{q} \lambda_8 q = \delta m \left(\frac{\bar{p}'p' + \bar{n}'n - 2\bar{\Lambda}'\Lambda'}{\sqrt{3}} \right)$$

leaves only the p' and n' masses degenerate.

$$m_{p'} = m_{n'} = m - \frac{\delta m}{\sqrt{3}} \equiv m_{N'}$$

$$m_{\Lambda'} = m + \frac{2}{\sqrt{3}} \delta m$$

$$\delta m = \sqrt{3}\,(m_{\Lambda'} - m_{N'})$$

The transformation $q \longrightarrow (1 + i\varepsilon_\alpha \frac{\lambda_\alpha}{2}) q$ leads to

$$\delta m\, \bar{q}\lambda_8 q \longrightarrow \delta m\, \bar{q}(1 - i\varepsilon_\alpha \tfrac{\lambda_\alpha}{2})\lambda_8 (1 + i\varepsilon_\alpha \tfrac{\lambda_\alpha}{2}) q$$
$$= \delta m\, \bar{q}\lambda_8 q + \tfrac{i}{4}\delta m\, \varepsilon_\alpha \bar{q}[\lambda_8, \lambda_\alpha] q$$

$$\partial_\mu j^\alpha_\mu = i\, \frac{\delta m}{4} \bar{q}[\lambda_8, \lambda_\alpha] q$$

$$[\lambda_8, \lambda_\alpha] \begin{cases} = 0 & \text{for } \alpha = 1,2,3,8 \\ \neq 0 & \text{for } \alpha = 4,5,6,7 \end{cases}$$

With the mass difference term the strangeness-changing currents are no longer conserved, whereas the isospin and hypercharge currents are unaffected. Likewise for the axial vector currents we get

$$\partial_\mu j^\alpha_{5\mu} = \mathcal{O}(m) + \mathcal{O}(\delta m).$$

It is convenient to talk about the j^α_μ and $j^\alpha_{5\mu}$ interacting linearly with "external fields" - objects which are outside the hadronic world (e.g. the electromagnetic field or the lepton current, $\bar{e}\gamma_\mu(1+\gamma_5)\nu$). Or we may think of these fictitious fields as "probes" of the hadronic currents.

$$\mathcal{L}_{INT} = W^\gamma_\mu j^\gamma_\mu + W^\gamma_{5\mu} j^\gamma_{5\mu}$$

Consider again $q \longrightarrow (1 + i\varepsilon_\alpha \frac{\lambda_\alpha}{2})q$.

$$\begin{Bmatrix} j^\gamma_\mu \\ j^\gamma_{5\mu} \end{Bmatrix} \longrightarrow i\bar{q}(1 - i\varepsilon_\alpha \tfrac{\lambda_\alpha}{2}) \begin{Bmatrix} \gamma_\mu \\ \gamma_\mu\gamma_5 \end{Bmatrix} \frac{\lambda_\gamma}{2} (1 + i\varepsilon_\alpha \tfrac{\lambda_\alpha}{2})q$$

$$= \begin{Bmatrix} j^\gamma_\mu \\ j^\gamma_{5\mu} \end{Bmatrix} - \frac{\varepsilon_\alpha}{4}\bar{q} \begin{Bmatrix} \gamma_\mu \\ \gamma_\mu\gamma_5 \end{Bmatrix} [\lambda_\gamma, \lambda_\alpha] q$$

$$= \begin{Bmatrix} j^\gamma_\mu \\ j^\gamma_{5\mu} \end{Bmatrix} - \varepsilon_\alpha f_{\alpha\beta\gamma} \begin{Bmatrix} j^\beta_\mu \\ j^\beta_{5\mu} \end{Bmatrix}$$

So the Gell-Mann-Levy equation now reads

$$\partial_\mu j^\alpha_\mu = \mathcal{O}(\delta m) + f_{\alpha\beta\gamma} j^\beta_\mu W^\gamma_\mu + f_{\alpha\beta\gamma} j^\beta_{5\mu} W^\gamma_{5\mu}$$

(where the $\mathcal{O}(\delta m)$ term is due to the medium-strong mass difference, as before)

Similarly

$$q \longrightarrow (1 + i\varepsilon_\alpha \gamma_5 \frac{\lambda_\alpha}{2}) q$$

gives

$$\partial_\mu j^\alpha_{5\mu} = O(m) + O(\delta m) + f_{\alpha\beta\gamma} j^\beta_{5\mu} W^\gamma_\mu + f_{\alpha\beta\gamma} j^\beta_\mu W^\gamma_{5\mu} .$$

These equations are known as the divergence conditions or as Veltman's equations (Veltman '66). (Special cases were used earlier by Schwinger, Adler and others.)

Let us now apply this formalism to the electromagnetic and weak interactions:

(i) Electromagnetic interaction: $eA_\mu \, (J_\mu{}^3 + J_\mu{}^8/\sqrt{3})$.
Set $W_\mu{}^3 = eA_\mu$, $W_\mu{}^8 = eA_\mu/\sqrt{3}$, and all others = 0.
From Veltman's equation and $f_{123} = 1$, $f_{128} = 0$, it is easy to show that

$$\partial_\mu j_\mu^{1\pm i2} = \mp i e A_\mu j_\mu^{1\pm i2},$$

which is expected from the gauge derivative substitution

$$\partial_\mu \longrightarrow \partial_\mu \pm i e A_\mu .$$

Note that the 1st and 2nd components of the isospin are no longer conserved.

(ii) Semileptonic weak interaction:

$$\frac{G}{\sqrt{2}} \left[i\bar{e}\gamma_\mu(1+\gamma_5)\nu + i\bar{\mu}\gamma_\mu(1+\gamma_5)\nu' \right] \left[\cos\theta \, (j_\mu^{1+i2} + j_{5\mu}^{1+i2}) + \sin\theta \, (j_\mu^{4+i5} + j_{5\mu}^{4+i5}) \right] + H.C.$$

We can identify

$$\begin{Bmatrix} W_\mu^1 \\ W_\mu^4 \end{Bmatrix} = \frac{G}{\sqrt{2}} \begin{Bmatrix} \cos\theta \\ \sin\theta \end{Bmatrix} \left[i\bar{e}\gamma_\mu(1+\gamma_5)\nu + i\bar{\nu}\gamma_\mu(1+\gamma_5)e \right] + \{\mu \leftrightarrow e, \nu' \leftrightarrow \nu\}$$

$$\begin{Bmatrix} W_\mu^2 \\ W_\mu^5 \end{Bmatrix} = \frac{iG}{\sqrt{2}} \begin{Bmatrix} \cos\theta \\ \sin\theta \end{Bmatrix} \left[i\bar{e}\gamma_\mu(1+\gamma_5)\nu - i\bar{\nu}\gamma_\mu(1+\gamma_5)e \right] + \{\mu \leftrightarrow e, \nu' \leftrightarrow \nu\}$$

and similarly for the $W_{5\mu}$'s. When there are semileptonic weak interactions (e.g. $n \rightarrow p + e^- + \bar{\nu}$), one finds from Veltman's equation

$$\partial_\mu \left(j_\mu^3 + \frac{1}{\sqrt{3}} j_\mu^8 \right) \neq 0 .$$

This is not surprising since what is conserved is the <u>total</u> electric charge current of the hadrons <u>and</u> leptons. In fact it is easy to show

$$\partial_\mu \left(j_\mu^3 + \frac{1}{\sqrt{3}} j_\mu^8 \right) = \partial_\mu \left(i\bar{e}\gamma_\mu e + i\bar{\mu}\gamma_\mu \mu \right) .$$

(Hint: Consider $\begin{Bmatrix} e \\ \mu \end{Bmatrix} \longrightarrow (1 - i\varepsilon(x)) \begin{Bmatrix} e \\ \mu \end{Bmatrix}$ and use the Gell-Mann-Lévy equation.)

<u>Derivation of Current Commutation Relations (CCR)</u>

Use the covariant form of Heisenberg's equation,

$$\partial_\mu j_\mu^\alpha = i\left[j_\mu^\alpha(x) , P_\mu(x_0) \right]$$

$$\partial_\mu j_{5\mu}^\alpha = i\left[j_{5\mu}^\alpha(x) , P_\mu(x_0) \right] ,$$

where P_μ is the total energy-momentum operator

$$P_\mu = -i \int d^3x \, T_{4\mu}$$

$$T_{\mu\nu} = -\frac{\partial \mathcal{L}}{\partial(\partial_\mu \psi_r)} \partial_\nu \psi_r + \delta_{\mu\nu} \mathcal{L}$$

Look now at that part of P_μ which is due to

$$\mathcal{L}_{INT} = W_\mu^\gamma j_\mu^\gamma + W_{5\mu}^\gamma j_{5\mu}^\gamma .$$

Call it $P_\mu^{(int)}$. Because $\mathcal{L}_{int}$ involves no derivatives,

$$T_{\mu\nu}^{(INT)} = \delta_{\mu\nu} \mathcal{L}_{INT} ,$$

and we get

$$P_4^{(int)} = -i \int d^3 x \mathcal{L}_{int} , \qquad \vec{P}^{(int)} = 0.$$

(This is no surprise, since $\mathcal{L}_{int} = -H_{int}$ when it involves no derivatives.)

$$\partial_\mu j_\mu^\gamma = \mathcal{O}(\delta m) + [\, j_4^\gamma , \int d^3x \, (W_\mu^\gamma j_\mu^\gamma + W_{5\mu}^\gamma j_{5\mu}^\gamma)]$$

The $\mathcal{O}(\delta m)$ term is due to the medium-strong part of $P_4^{(int)}$. Compare this with the expression we got from the G-M-L equation.

$$\partial_\mu j_\mu^\alpha = \mathcal{O}(\delta m) + f_{\alpha\beta\gamma} j_\mu^\beta W_\mu^\gamma + f_{\alpha\beta\gamma} j_{5\mu}^\beta W_{5\mu}^\gamma$$

We use W to probe the hadronic world. So assuming it can be controlled, set $W_{5\mu}^\gamma = 0$ for every γ, $W_\mu^\gamma \neq 0$ for a particular γ and assume W_μ^γ uniform in space at a given time. Hence with

$$\int d^3x \, W_\mu^\gamma j_\mu^\gamma = W_\mu^\gamma \int d^3x \, j_\mu^\gamma$$

we get

$$[j_0^\alpha, \int d^3x\, j_\mu^\gamma] = -i f_{\alpha\beta\gamma} j_\mu^\beta .$$

$(\beta \leftrightarrow \gamma)$

$$[j_0^\alpha(x), \int d^3x' j_\mu^\beta(x')]_{x_0'=x_0} = i f_{\alpha\beta\gamma} j_\mu^\gamma(x)$$

Set μ=o.

$$[j_0^\alpha, \int d^3x\, j_0^\beta] = i f_{\alpha\beta\gamma} j_0^\gamma$$

By covariance

$$[j_\mu^\alpha, \int d^3x\, j_0^\beta] = i f_{\alpha\beta\gamma} j_\mu^\gamma ,$$

$$[j_\mu^\alpha(x), Q^\beta(x_0)] = i f_{\alpha\beta\gamma} j_\mu^\gamma(x) .$$

A similar thing can be done for $\partial_\mu J_{5\mu}^\alpha$.

To summarize, $\partial_\mu J_\mu^\alpha$ and $\partial_\mu J_{5\mu}^\alpha$ are computed in two different ways: (i) the Gell-Mann-Lévy variational principle, and (ii) the Heisenberg equation. By demanding that the two approaches be consistent, we get equal time commutation relations of the form:

$$[j_\mu^\alpha(x), Q^\beta(x_0)] = i f_{\alpha\beta\gamma} j_\mu^\gamma(x)$$

$$[j_\mu^\alpha(x), Q_5^\beta(x_0)] = i f_{\alpha\beta\gamma} j_{5\mu}^\gamma(x)$$

$$[j_{5\mu}^\alpha(x), Q^\beta(x_0)] = i f_{\alpha\beta\gamma} j_{5\mu}^\gamma(x)$$

$$[j_{5\mu}^\alpha(x), Q_5^\beta(x_0)] = i f_{\alpha\beta\gamma} j_\mu^\gamma(x) .$$

These are the famous charge-current commutations of Gell-Mann ('62). The 1st and 3rd equations are old, since they merely specify how J^{α}_{μ} and $J^{\alpha}_{5\mu}$ transform under SU(3), the 2nd is new, and the 4th is quite nontrivial. These equations are called "once-integrated CCR."

Density Commutation Relations

We now show that these equations and the "unintegrated CCR" can be obtained directly from anticommutation relations among quark fields.

$$\{q^{+}_{\tau}(x), q_{\tau'}(x')\}_{x_0=x_0'} = \delta_{\tau\tau'}\,\delta^{(3)}(\vec{x}-\vec{x}') \qquad (\tau,\tau'=1,\cdots,12)$$

$$\{q_{\tau}(x), q_{\tau'}(x')\}_{x_0=x_0'} = \{q^{+}_{\tau}(x), q^{+}_{\tau'}(x')\}_{x_0=x_0'} = 0$$

In particular we are interested in commutators of the form

$$[\bar{q}(x)\,\mathcal{O}\,q(x), \bar{q}(x')\,\mathcal{O}'q(x')]_{x_0=x_0'} \qquad (\bar{q}=q^{+}\gamma_4),$$

which we calculate with the aid of two useful identities:

(i) $[AB, CD] \equiv -AC\{D,B\} + A\{C,B\}D - C\{D,A\}B + \{C,A\}DB$

(ii) $[\Gamma_a\lambda_\alpha, \Gamma_b\lambda_\beta] \equiv \frac{1}{2}\{\Gamma_a,\Gamma_b\}[\lambda_\alpha,\lambda_\beta] + \frac{1}{2}[\Gamma_a,\Gamma_b]\{\lambda_\alpha,\lambda_\beta\}$

where Γ_a, Γ_b are any of the 16 Dirac matrices.

So using (i) and the anticommutation relations, the commutator becomes

$$[q^\dagger_\sigma(x)q_\tau(x), q^\dagger_{\sigma'}(x')q_{\tau'}(x')]_{x_0=x_0'}(\gamma_4\mathcal{O})_{\sigma\tau}(\gamma_4\mathcal{O}')_{\sigma'\tau'} =$$

$$= \left(q^\dagger_\sigma(x)\delta_{\sigma'\tau}q_{\tau'}(x') - q^\dagger_{\sigma'}(x')\delta_{\sigma\tau'}q_\tau(x)\right)_{x_0=x_0'}\delta^{(3)}(\vec{x}-\vec{x}')(\gamma_4\mathcal{O})_{\sigma\tau}(\gamma_4\mathcal{O}')_{\sigma'\tau'}$$

$$= q^\dagger(x)[\gamma_4\mathcal{O}, \gamma_4\mathcal{O}']q(x')\,\delta^{(3)}(\vec{x}-\vec{x}')\Big|_{x_0=x_0'}.$$

The appearance of the delta function is not surprising, since causality requires that at space-like separations the densities must commute. Now set $\gamma_4\mathcal{O} = \Gamma_a\frac{\lambda_\alpha}{2}$, $\gamma_4\mathcal{O}' = \Gamma_b\frac{\lambda_\beta}{2}$ and use identity (ii). For example,

$$[i\bar{q}\gamma_\mu\tfrac{\lambda_\alpha}{2}q, i\bar{q}\gamma_4\tfrac{\lambda_\beta}{2}q] =$$

$$= -\frac{1}{8}q^\dagger(x)\Big(\underbrace{\{\gamma_4\gamma_\mu, 1\}}_{2\gamma_4\gamma_\mu}\underbrace{[\lambda_\alpha,\lambda_\beta]}_{2if_{\alpha\beta\gamma}\lambda_\gamma} + \underbrace{[\gamma_4\gamma_\mu, 1]}_{0}\{\lambda_\alpha,\lambda_\beta\}\Big)q(x')\delta^{(3)}(\vec{x}-\vec{x}')$$

$$= -if_{\alpha\beta\gamma}\bar{q}(x)\frac{\lambda_\gamma}{2}q(x')\delta^{(3)}(\vec{x}-\vec{x}')$$

corresponds to the "unintegrated CCR,"

$$[j^\alpha_\mu(x), j^\beta_4(x')]_{x_0=x_0'} = -f_{\alpha\beta\gamma}j^\gamma_\mu(x)\,\delta^{(3)}(\vec{x}-\vec{x}').$$

Taking the space integral, we get the familiar once-integrated CCR,

$$[j^\alpha_\mu(x), \int d^3x'\,j^\beta_0(\vec{x}', x_0)] = [j^\alpha_\mu(x), Q^\beta(x_0)] = if_{\alpha\beta\gamma}j^\gamma_\mu(x).$$

(The converse, however, is not true; the once-integrated CCR does not imply the unintegrated CCR.) The other three CCR's are obtained in a similar manner; e.g. for the axial-axial CCR use

$$\underbrace{\{\gamma_4\gamma_\mu\gamma_5,\gamma_5\}}_{2\gamma_4\gamma_\mu}\underbrace{[\lambda_\alpha,\lambda_\beta]}_{2if_{\alpha\beta\gamma}\lambda_\gamma}+\underbrace{[\gamma_4\gamma_\mu\gamma_5,\gamma_5]}_{0}\{\lambda_\alpha,\lambda_\beta\}$$

to get

$$\left[j^\alpha_{5\mu}(x),j^\beta_{54}(x')\right]_{x_0=x_0'}=-f_{\alpha\beta\gamma}j^\gamma_\mu(x)\,\delta^{(3)}(\vec{x}'-\vec{x}).$$

It is also possible to get new relations if you take the quark model seriously. (Gell-Mann '64; Feynman, Gell-Mann and Zweig '64.) Consider the CR of the space-space components of the currents, $[j^\alpha_k, j^\beta_\ell]$. We must work with

$$\{\gamma_4\gamma_k,\gamma_4\gamma_\ell\}[\lambda_\alpha,\lambda_\beta]+[\gamma_4\gamma_k,\gamma_4\gamma_\ell]\{\lambda_\alpha,\lambda_\beta\}=$$

$$=-2\delta_{k\ell}\,2if_{\alpha\beta\gamma}\lambda_\gamma+2\gamma_4\varepsilon_{k\ell m}\gamma_m\gamma_5\left(\tfrac{4}{3}\delta_{\alpha\beta}+2d_{\alpha\beta\gamma}\lambda_\gamma\right),$$

where we have used

$$\{\gamma_4\gamma_k,\gamma_4\gamma_\ell\}=-\{\gamma_k,\gamma_\ell\}=-2\delta_{k\ell}$$

$$[\gamma_4\gamma_k,\gamma_4\gamma_\ell]=\gamma_4(\gamma_k\gamma_4\gamma_\ell-\gamma_\ell\gamma_4\gamma_k)=$$

$$=-\gamma_4(\varepsilon_{k\ell m}\gamma_m\gamma_5-\varepsilon_{\ell k m}\gamma_m\gamma_5)=-2\gamma_4\varepsilon_{k\ell m}\gamma_m\gamma_5.$$

So we get

$$[j^\alpha_k(x),j^\beta_\ell(x')]_{x_0=x_0'}=$$

$$=\delta^{(3)}(\vec{x}-\vec{x}')\left\{\delta_{k\ell}f_{\alpha\beta\gamma}j^\gamma_4(x)-i\varepsilon_{k\ell m}\left[\sqrt{\tfrac{2}{3}}\delta_{\alpha\beta}j^0_{5m}(x)+d_{\alpha\beta\gamma}j^\gamma_{5m}(x)\right]\right\}$$

which has both f and d structure constants. In general, if $[\Gamma_a,\Gamma_b]\neq 0$, we get structure constants going like $\delta_{\alpha\beta}$ and $d_{\alpha\beta\gamma}$ on the RHS, whereas $\{\Gamma_a,\Gamma_b\}\neq 0$ gives $f_{\alpha\beta\gamma}$. Unfortunately, there is as yet no experimental verification of the space-space CCR. Another example involving d-type structure constants is

$$[i\bar{q}\gamma_5 \frac{\lambda_\alpha}{2} q, i\bar{q}\gamma_4\gamma_5 \frac{\lambda_\beta}{2} q] = -\left(\frac{1}{3}\delta_{\alpha\beta}\bar{q}q + d_{\alpha\beta\gamma}\bar{q}\frac{\lambda_\gamma}{2}q\right),$$

where we have used

$$\underbrace{\{\gamma_4\gamma_5, \gamma_5\}}_{0}[\lambda_\alpha, \lambda_\beta] + \underbrace{[\gamma_4\gamma_5, \gamma_5]}_{2\gamma_4}\{\lambda_\alpha, \lambda_\beta\} = 2\gamma_4\left(\frac{4}{3}\delta_{\alpha\beta} + 2d_{\alpha\beta\gamma}\lambda_\gamma\right).$$

In general

$$[(\bar{q}\mathcal{O}^{(a)}q)_x, (\bar{q}\mathcal{O}^{(b)}q)_{x'}]_{x_0=x_0'} = C_{abc}(\bar{q}\mathcal{O}^{(c)}q)_x \delta^{(3)}(\vec{x}-\vec{x}')$$

is a commutation relation between any pair of 16 x (8 + 1) = 144 current densities. In practical applications only a few of the commutation relations have been tested. The ones we will take seriously are the time-time and space-time commutation relations of vector and axial vector densities; when integrated once these lead to the current-charge commutation relations derived earlier using a different method. Other relations are often highly model-dependent in the sense that they are true only in the quark model. Later in the course we mention the gauge field algebra, which requires the vanishing of the space-space commutation relations:

$$[j_k^\alpha(x), j_\ell^\beta(x')] = [j_k^\alpha(x), j_{5\ell}^\beta(x')] = [j_{5k}^\alpha(x), j_{5\ell}^\beta(x')] = 0$$

where the j's are now proportional to vector and axial-vector gauge fields.

Schwinger term

Unintegrated densities are highly singular objects. In fact, commutation relations among them involve delta functions, and some of the mathematical manipulations might not be justified. According to the CCR which we have derived,

$$[j_0^\alpha(x),\, j_k^\alpha(x')]_{x_0=x_0'} = 0 \quad \text{because} \quad f_{\alpha\alpha\gamma} = 0 .$$

Schwinger ('59) was able to show that this relation cannot be correct. Consider the case where j_μ^α is conserved (e.g. isospin or the electric charge, $3 + 8/\sqrt{3}$ th component).

$$[j_0^\alpha(0,0),\, j_k^\alpha(\vec{x},0)] = 0$$

$$[j_0^\alpha(0,0),\, \vec{\nabla}\cdot\vec{j}^\alpha(\vec{x},0)] = -[j_0^\alpha(0,0),\, \tfrac{\partial}{\partial t} j_0^\alpha(\vec{x},0)] = 0$$

Take $\vec{x} = 0$.

$$\begin{aligned} 0 &= -i\langle 0|[j_0^\alpha(0),\, \tfrac{\partial}{\partial t} j_0^\alpha(0)]|0\rangle = \\ &= \langle 0|[j_0^\alpha(0), [H, j_0^\alpha(0)]]|0\rangle = \\ &= \sum_n \langle 0|j_0^\alpha(0)|n\rangle\langle n|[H, j_0^\alpha(0)]|0\rangle - \sum_n \langle 0|[H, j_0^\alpha(0)]|n\rangle\langle n|j_0^\alpha(0)|0\rangle \\ &= 2\sum_n E_n |\langle 0|j_0^\alpha(0)|n\rangle|^2 \end{aligned}$$

The last expression is positive definite since $E_n \geq 0$. Hence there is a contradiction.

It turns out that, if we are more careful and use $\lim_{\epsilon\to 0} \bar{q}(\vec{x}+\vec{\epsilon}, x_0)\, \Theta\, q(\vec{x}-\vec{\epsilon}, x_0)$ as the definition of our currents, then we pick up an additional term in our CCR:

$$[j_0^\alpha(\vec{x},0), j_k^\beta(0)] = i f_{\alpha\beta\gamma} j_k^\gamma(\vec{x},0)\,\delta^{(3)}(\vec{x}) - i\, C_{\alpha\beta}\, \partial_k \delta^{(3)}(\vec{x})$$

+ (possible higher derivatives)

where, unfortunately, $c_{\alpha\alpha}$ goes to infinity like $1/\varepsilon^2$ as $\vec{\varepsilon} \to 0$

This extra term (called the Schwinger term) does not affect the once-integrated CCR (i.e. current-charge CR) because $\vec{\nabla}\delta^{(3)}(\vec{x})$ vanishes when integrated.

Historically the necessity of the Schwinger term was first pointed out by Goto and Imamura ('55) who looked at Källén's representation for $\langle 0|[j_\mu, j_\nu]|0\rangle$ in QED:

$$\langle 0|[j_\mu(x), j_\nu(x')]|0\rangle = i\int_0^\infty dm^2\, \rho(m^2)\left(\delta_{\mu\nu} - \frac{\partial_\mu \partial_\nu}{m^2}\right)\Delta(x-x'; m^2)$$

where

$$\Delta(x-x'; m^2) = \frac{-i}{(2\pi)^3}\int \frac{d^3p}{2p_0}\left(e^{ip\cdot(x-x')} - e^{-ip\cdot(x-x')}\right)$$
$$\left(p_0 = \sqrt{|\vec{p}|^2 + m^2}\right)$$

and

$$\left(\delta_{\mu\nu} - \frac{p_\mu p_\nu}{p^2}\right)\rho(-p^2) = (2\pi)^3 \sum_n \delta^{(4)}(p-p_n)\langle 0|j_\mu(0)|n\rangle\langle n|j_\nu(0)|0\rangle$$

(This representation can be obtained by inserting a complete set of states in $\langle 0|[j_\mu, j_\nu]|0\rangle$ and applying translational invariance.). Set $x_0 = x_0'$, $\mu = 4$, $\nu = k$ and use the well-known property of the invariant function,

$$\left.\frac{\partial}{\partial x_0}\Delta(x-x'; m^2)\right|_{x_0 = x_0'} = -\delta^{(3)}(\vec{x}-\vec{x}').$$

We then obtain

$$\langle 0|[j_4(x), \vec{j}(x')]|0\rangle = \vec{\nabla}\delta^{(3)}(\vec{x}-\vec{x}')\int dm^2 \frac{\rho(m^2)}{m^2}$$

or, equivalently

$$\langle 0|[j_0(\vec{x},0), j_k(0)]|0\rangle = -i\,\partial_k \delta^{(3)}(\vec{x})\int dm^2 \frac{\rho(m^2)}{m^2} \quad .$$

Since $\rho(m^2)$ is essentially the probability for an external electromagnetic field to create an object of total mass m, the integral is positive definite. But we can identify the above expression with $\langle 0|$Schwinger term$|0\rangle$. This argument shows that the Schwinger term cannot vanish if the theory is not to be completely trivial. We may add that the coefficients $c_{\alpha\beta}$ are model dependent, and, in general, are operators. (We will later show, however, that in the gauge field algebra $c_{\alpha\beta}$ is a finite c-number.) At this stage one thing is clear; we need Schwinger terms if the theory is to be consistent with Lorentz invariance and the positive definiteness of probability.

The Schwinger disease does not have much to do with conserved currents even though we deduced a contradiction only for a conserved current. In the Goto-Imamura approach, if the current is not conserved, there would be two independent spectral functions (spin 1 and spin 0), but the argument for the necessity of the Schwinger term would remain unchanged (Okubo '66). Later in the course we discuss to what extent we can "measure" the Schwinger terms.

Charge-Charge Commutation Relations and Chiral SU(2) ⊗ SU(2)

So far we have talked about (once-integrated) charge-current CR's and (unintegrated) current-current CR's, but historically the oldest form is the twice integrated charge-charge CR. Isospin was first discussed in analogy with angular momentum:

$$[T_\alpha, T_\beta] = i\,\varepsilon_{\alpha\beta\gamma} T_\gamma \qquad \alpha,\beta,\gamma = 1,2,3$$

Generalize to SU(3):

$$[Q^\alpha, Q^\beta] = i f_{\alpha\beta\gamma} Q^\gamma \qquad \alpha,\beta,\gamma = 1,\cdots,8.$$

This can also be obtained from $[j^\alpha_\mu(x), Q^\beta(x_0)] = i f_{\alpha\beta\gamma} j^\gamma_\mu(x)$ by setting $\mu = 0$ and integrating. Doing the same thing for the other three current-charge CR's, we get

$$[Q^\alpha, Q^\beta_5] = i f_{\alpha\beta\gamma} Q^\gamma_5 ,$$

$$[Q^\alpha_5, Q^\beta_5] = i f_{\alpha\beta\gamma} Q^\gamma .$$

If we define

$$Q^\alpha_\pm = \tfrac{1}{2}(Q^\alpha \pm Q^\alpha_5),$$

then

$$[Q^\alpha_+, Q^\beta_-] = \tfrac{1}{4}[Q^\alpha + Q^\alpha_5, Q^\beta - Q^\beta_5]$$

$$= \tfrac{1}{4}\Big(\underbrace{[Q^\alpha, Q^\beta]}_{i f_{\alpha\beta\gamma} Q^\gamma} - \underbrace{[Q^\alpha, Q^\beta_5]}_{i f_{\alpha\beta\gamma} Q^\gamma_5} + \underbrace{[Q^\alpha_5, Q^\beta]}_{i f_{\alpha\beta\gamma} Q^\gamma_5} - \underbrace{[Q^\alpha_5, Q^\beta_5]}_{i f_{\alpha\beta\gamma} Q^\gamma}\Big)$$

$$= 0,$$

$$[Q_+^\alpha, Q_+^\beta] = i f_{\alpha\beta\gamma} Q_+^\gamma ,$$
$$[Q_-^\alpha, Q_-^\beta] = i f_{\alpha\beta\gamma} Q_-^\gamma .$$

Thus we have two independent SU(3) commutation relations usually called chiral SU(3) ⊗ SU(3) (to distinguish it from collinear SU(3) ⊗ SU(3), which we will not discuss in this course). Note that $Q_\pm^\alpha$ is not a constant of the motion because Q_5^α is not. Chiral SU(3) ⊗ SU(3) ≡ (transf. on right-handed q) ⊗ (transf. on left-handed q).

$$q_L = \tfrac{1}{2}(1+\gamma_5)q \qquad \bar{q}_L = \tfrac{1}{2}\bar{q}(1-\gamma_5)$$
$$q_R = \tfrac{1}{2}(1+\gamma_5)q \qquad \bar{q}_R = \tfrac{1}{2}\bar{q}(1+\gamma_5)$$

$Q_\pm^\alpha$ are the space integrals of the zeroth components of currents generated by chiral SU(3) transformations. More precisely, the left-handed SU(3) transformation,

$$q \longrightarrow \left(1 + i\,\varepsilon_\alpha \frac{(1+\gamma_5)}{2}\frac{\lambda_\alpha}{2}\right) q$$

or, equivalently,

$$q_L \longrightarrow \left(\frac{1+\gamma_5}{2} + \left[\frac{1+\gamma_5}{2}\right]^2 i\,\varepsilon_\alpha \frac{\lambda_\alpha}{2}\right) q = \left(1 + i\,\varepsilon_\alpha \frac{\lambda_\alpha}{2}\right) q_L$$

$$\bar{q}_L \longrightarrow \bar{q}_L \left(1 - i\,\varepsilon_\alpha \frac{\lambda_\alpha}{2}\right)$$

$$q_R \longrightarrow \left(\frac{1-\gamma_5}{2} + \left[\frac{1-\gamma_5}{2}\right]\left[\frac{1-\gamma_5}{2}\right] i\,\varepsilon_\alpha \frac{\lambda_\alpha}{2}\right) q = q_R$$

$$\bar{q}_R \longrightarrow \bar{q}_R$$

generates the current corresponding to Q_+^α ,

$$i\,\bar{q}_L \gamma_\mu \frac{\lambda_\alpha}{2} q_L = i\,\bar{q}\,\gamma_\mu \frac{(1+\gamma_5)}{2}\frac{\lambda_\alpha}{2} q .$$

Similarly, the right-handed SU(3) transformation,

$$q \to \left(1 + i\varepsilon_\alpha \left(\frac{1-\gamma_5}{2}\right)\frac{\lambda_\alpha}{2}\right)q$$

$$q_R \to \left(1 + i\varepsilon_\alpha \frac{\lambda_\alpha}{2}\right)q_R \qquad q_L \to q_L$$

$$\bar{q}_R \to \bar{q}_R\left(1 - i\varepsilon_\alpha \frac{\lambda_\alpha}{2}\right) \qquad \bar{q}_L \to \bar{q}_L,$$

generates the current corresponding to Q^α_-,

$$i\bar{q}_R\gamma_\mu \frac{\lambda_\alpha}{2} q_R = i\bar{q}\gamma_\mu\left(\frac{1-\gamma_5}{2}\right)\frac{\lambda_\alpha}{2}q.$$

In the free quark Lagrangian, $\mathcal{L} = -\bar{q}(\gamma_\mu\partial_\mu + m)q$, the kinetic term

$$-\bar{q}\gamma_\mu\partial_\mu q = -\bar{q}_L\gamma_\mu\partial_\mu q_L - \bar{q}_R\gamma_\mu\partial_\mu q_R$$

transforms like (1, 1) [(1,1) ≡ (RH-SU(3) singlet, LH-SU(3) singlet)] under chiral SU(3) ⊗ SU(3). In other words, the kinetic term is invariant under the chiral SU(3) transformations,

$q \to \left(1 + i\varepsilon_\alpha\left(\frac{1\pm\gamma_5}{2}\right)\frac{\lambda_\alpha}{2}\right)q$, which generate the currents corresponding to $Q^\alpha_\pm$. In contrast the mass term

$$-m\bar{q}q = -m(\bar{q}_R q_L + \bar{q}_L q_R)$$

transforms like $(3^*,3) + (3,3^*)$. It is not invariant under the above transformations and therefore breaks the chiral symmetry.

Can we still hope that chiral symmetry has something to do with reality? The classification of states attempted by Gell-Mann ('64) and others has not been so successful. In this scheme there is a doubling of meson states, (1,8) + (8,1) or $(3^*,3) + (3,3^*)$; the existence of a pseudoscalar (vector) octet implies the existence of a scalar (axial) octet. It also requires a "ninth baryon," i.e., a $1/2^-$ unitary singlet (perhaps the $Y^*(1405)$?). Chiral symmetry was not so popular as a good symmetry until 1967 when the idea was revived in connection with Weinberg's mass formula, $m_{A_1} = \sqrt{2}\, m_\rho$. We will come back to this later (Chapter 6). Meanwhile, we take the point of view that chiral symmetry may be a badly broken symmetry but the commutation relations are exact. This is somewhat analogous to working with $[x_i, p_j] = i\delta_{ij}$, a relation which is supposed to be exact even though x_i and p_j are not in general constants of the motion. The commutation relation is very useful and is indeed the whole of nonrelativistic quantum mechanics.

Currents Appearing in Electromagnetic and Semileptonic Processes

Treating the strong interactions to all orders and the e.m. and weak interactions to first order, we split off the nonstrong interactions from the rest of the Hamiltonian.

$$H_{tot} = (H_{free} + H_{STRONG}) + H_{ext}$$

For e.m. and semileptonic interactions, we can write

$$H_{ext} = -\int d^3x\, \mathcal{L}_{INT}^{(ext)} = -\int d^3x \left(W_\mu^\alpha j_\mu^\alpha + W_{5\mu}^\alpha j_{5\mu}^\alpha \right).$$

In the absence of the external interaction, the S-matrix is

$$S_{BA} = {}_{out}\langle B|A\rangle_{IN} .$$

But in the presence of the external interactions, it becomes

$$_{out}\langle B|A\rangle_{IN} \xrightarrow{\text{ext. INT.}} {}_{out}\langle B|A\rangle_{IN} - i\,{}_{OUT}\langle B|\int dt\, H_{ext}|A\rangle_{IN} + \cdots$$

(For an elementary derivation, see e.g. Nishijima's book, pp. 192-193.) Our treatment differs from the "interaction picture" in which the strong interaction is included with the external interaction $\left[H_{tot} = H_{free} + (H_{strong} + H_{ext})\right]$. The difference is merely a matter of grouping. In our picture the S-matrix modifying term

$$- i\,{}_{OUT}\langle B|\int dt\, H_{ext}|A\rangle_{IN} = i\int d^4x\, W_\mu^\gamma(x)\langle B|j_\mu^\gamma|A\rangle + \{\mu \leftrightarrow 5\mu\}$$

leads to the study of hadronic matrix elements of the form:

$\langle B|j_\mu^3(x) + \frac{1}{\sqrt{3}} j_\mu^8(x)|A\rangle$ in e.m. interactions,

$\langle B|j_\mu^{1\pm i2}(x)|A\rangle + \{\mu \leftrightarrow 5\mu\}$ in $\Delta S=0$ semileptonic int.,

$\langle B|j_\mu^{4\pm i5}(x)|A\rangle + \{\mu \leftrightarrow 5\mu\}$ in $\Delta Q = \Delta S = \pm 1$ " " .

(When $|A\rangle$ and $|B\rangle$ are not single-particle states, they are understood to be $|A\rangle_{in}$ and $|B\rangle_{out}$.) If hadrons are made up of quarks, then

$$j^{\alpha}_{\mu} = i\bar{q}\gamma_{\mu}\frac{\lambda_{\alpha}}{2}q \quad \text{and} \quad j^{\alpha}_{5\mu} = i\bar{q}\gamma_{\mu}\gamma_{5}\frac{\lambda_{\alpha}}{2}q.$$

If there are no fundamental fields, then the matrix elements appearing in e.m. and semileptonic processes can be taken as definitions of j^{α}_{μ} and $j^{\alpha}_{5\mu}$. In any case, assume that these currents satisfy charge-current commutation relations abstracted from the free quark model.

As an example of the way in which the e.m. interaction can be used to explore the matrix elements of j^{3}_{μ} and j^{8}_{μ}, consider photo- or electroproduction processes of the form

real photon (photoprod.) $q^2=(p_B-p_A)^2=0$

virtual photon (electroprod.) $q^2 \neq 0$.

This measures $\langle B|(j^{3}_{\mu}+\frac{1}{\sqrt{3}}j^{8}_{\mu})|A\rangle$. As another example, consider $\rho^{0} \rightarrow e^{+} + e^{-}$. The initial and final states are given by

$$|A\rangle = |\rho^{0}\rangle,$$
$$|B\rangle = |0\rangle.$$

This process

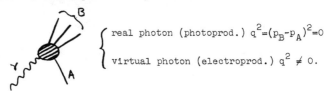

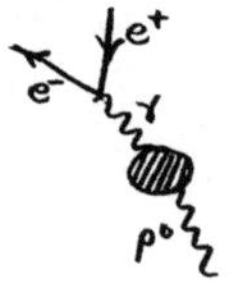

measures the matrix element $\langle 0|j^{3}_{\mu}|\rho^{0}\rangle$; note that j^{8}_{μ} gives no contribution because of isospin or G-parity.

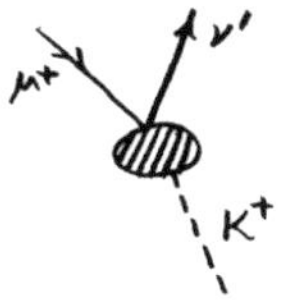

Turning now to the weak interactions, we may consider $K^+_{\mu 2}$ decay which measures $\langle 0 | j^{4-i5}_{5\mu} | K^+ \rangle$. Only the axial current contributes because of parity. (In the rest system, K^+ is odd and the vacuum is even under parity; j^{4-i5}_{54} is odd but j^{4-i5}_{4} is even.)

As a consequence of translational invariance, any operator can be written as

$$\mathcal{O}(x) = e^{-iP \cdot x} \mathcal{O}(0) e^{iP \cdot x}$$

where P is the total energy-momentum operator. To prove this, simply look at the solution to the differential (Heisenberg) equation

$$\partial_\mu \mathcal{O}(x) = i [\mathcal{O}(x), P_\mu]$$

subject to the boundary condition at x = 0. We get

$$\langle B | j^\alpha_\mu(x) | A \rangle = \langle B | j^\alpha_\mu(0) | A \rangle e^{i(p_A - p_B) \cdot x} .$$

Because we believe in translational invariance, it is sufficient to talk only about $\langle B | j^\alpha_\mu (0) | A \rangle$ and $\langle B | j^\alpha_{5\mu} (0) | A \rangle$.

Further examples of processes which measure the hadronic current matrix elements are:

(i) Electron-proton scattering

The matrix element of the isovector part of the electromagnetic current taken between two proton states is of the form

$$\langle p', s' | j_\mu^3 | p, s \rangle = \sqrt{\frac{m_p^2}{E'EV^2}}\, \frac{1}{2}\, \bar{u}' [i\gamma_\mu F_1^{(V)}(t) - i\sigma_{\mu\nu} q_\nu F_2^{(V)}(t)] u .$$

$$t = -(p'-p)^2 \quad , \quad q = p' - p$$

This is the most general form from parity, Lorentz covariance and current conservation. For instance, there is no $q_\mu \bar{u}'u$ term because current conservation implies $iq_\mu \langle B | j_\mu^3 | A \rangle = 0$ whereas $q_\mu (q_\mu \bar{u}'u) = q^2 \bar{u}'u \neq 0$. In contrast,

$$q_\mu \left\{ \begin{matrix} \bar{u}' \gamma_\mu u \\ q_\nu \bar{u}' \sigma_{\mu\nu} u \end{matrix} \right\} = 0 .$$

Also $(p'+p)_\mu\, \bar{u}'u$ is allowed, but this can be expressed as a linear combination of $i\bar{u}'\gamma_\mu u$ and $q_\nu \bar{u}' \sigma_{\mu\nu} u$.

(ii) π^+ decay

$$\langle 0 | j_{5\mu}^{1-i2}(0) | \pi^+ \rangle = i\sqrt{2}\, C_\pi\, p_\mu \frac{1}{\sqrt{2\omega V}}$$

since the only 4-vector available for a spin zero particle is p_μ . The $\sqrt{2\omega V}$ term comes from the wave function of the annihilated π^+ evaluated at $x = 0$, and the $\sqrt{2}$ is inserted so that

$$\langle 0 | j_{5\mu}^{\alpha}(0) | \pi^\alpha \rangle = \frac{i\, C_\pi\, p_\alpha}{\sqrt{2\omega V}} \qquad (\alpha \text{ not summed}).$$

Matrix Elements of Charge Operators; Ademollo-Gatto Theorem

By translational invariance ($Q = Q^\alpha$ or Q_5^α)

$$\langle B|Q|A\rangle = \langle B|\int d^3x\, j_0(x)|A\rangle = \int d^3x\, e^{i(p_A - p_B)\cdot x}\langle B|j_0(0)|A\rangle .$$

So at zero momentum transfer

$$\lim_{p_A \to p_B} \langle B|Q|A\rangle = V\langle B|j_0(0)|A\rangle .$$

The normalization volume V cancels with the 1/V coming from $\langle B|j_0(0)|A\rangle$, so that at zero momentum transfer the zeroth component of the currents appearing in electromagnetic and weak processes just measure the matrix elements of the charge operator. Between free protons, for example,

$$\langle p,s|Q^3|p,s\rangle = V\frac{m}{EV}\,\frac{1}{2}\,F_1^{(V)}(0)\,u^\dagger u .$$

Therefore

$$F_1^{(V)}(0) = 1 ,$$

where we have used $\langle p,s|Q^3|p,s\rangle = 1/2$ and $u^\dagger u = E/m$. (Note that our silly normalization factors are needed as long as we use $\langle p'|p\rangle = \delta_{pp'}$. An alternative approach is to use the covariant normalization convention $\langle p'|p\rangle = 2p_0\,\delta_{pp'}$.)

A note of caution; unlike $\langle Q^\alpha\rangle$, $\langle Q_5^\alpha\rangle$ is frame dependent.

$$\langle \vec{p},s|Q_5|\vec{p},s\rangle = \frac{m}{E}\,\frac{1}{2}\,F_A(0)\,u^\dagger(\vec{p},s)\gamma_5 u(\vec{p},s)$$

(where F_A is the axial-vector analog of $F_1^{(V)}$) becomes in the two-component notation

$$\langle \vec{p},s|Q_5|\vec{p},s\rangle = -\frac{m}{E}\,\frac{1}{2}F_A(0)\left(\frac{E+m}{2m}\right)\left(\frac{2\langle\vec{\sigma}\rangle\cdot\vec{p}}{E+m}\right)$$

$$= -\frac{1}{2}F_A(0)\,\frac{\langle\vec{\sigma}\rangle\cdot\vec{p}}{E} .$$

Now $\langle Q_5 \rangle = 0$ for a free proton at rest, but $\langle Q_5 \rangle \neq 0$ for a moving nucleon. In fact, it is easily seen that Q^α and Q_5^α have the same matrix elements for left-handed extremely relativistic protons ($\vec{p} \to \infty$).

Suppose a current is conserved. This means that its corresponding charge operator commutes with H, e.g. $[H^{strong}, Q^\alpha] = 0$. In the absence of electromagnetic interactions, Q^1 and Q^2 (as well as Q^3) are constants of the motion.

$$0 = \langle B | [H^{Strong}, Q^{1,2}] | A \rangle = (E_B - E_A) \langle B | Q^{1,2} | A \rangle$$

So

$$\langle B | Q^{1,2} | A \rangle = 0 \text{ unless } E_A = E_B .$$

When currents are conserved, the corresponding charge operators have no nonvanishing matrix elements between nondegenerate states; e.g. $Q^{1,2}$ connects n and p but does not connect n and N^{*+}

Look now at the commutation relation,

$$[Q^{1+i2}, Q^{1-i2}] = 2Q^3 \qquad (\text{use } f_{123} = 1),$$

between protons at rest (ignore spin).

$$2\langle p | Q^3 | p \rangle = \langle p | [Q^{1+i2}, Q^{1-i2}] | p \rangle$$

$$= \sum_m \langle p | Q^{1+i2} | m \rangle \langle m | Q^{1-i2} | p \rangle - \sum_m \langle p | Q^{1-i2} | m \rangle \langle m | Q^{1+i2} | p \rangle$$

Moreover, since Q^1 and Q^2 commute with H^{strong}, only the single neutron state contributes to the sum. Therefore

$$|\langle p | Q^{1+i2} | n \rangle|^2 = 2\langle p | Q^3 | p \rangle .$$

A similar argument for the pion gives

$$|\langle \pi^0 | Q^{1-i2} | \pi^+ \rangle|^2 = 2 \langle \pi^+ | Q^3 | \pi^+ \rangle$$
$$= 4 \langle p | Q^3 | p \rangle = 2 |\langle p | Q^{1+i2} | n \rangle|^2 .$$

Recalling that $\langle B | Q^{1 \pm i2} | A \rangle$ is measured in low momentum transfer β decay, we see that the matrix element for $\pi^+ \rightarrow \pi^0 + e^+ + \nu$ is related to the matrix element for $n \rightarrow p + e^- + \bar{\nu}$ by a simple numerical factor, viz. $\sqrt{2}$. This is the essence of the conserved vector theory (Feynman-Gell-Mann '58).

The situation is somewhat different, however, for S-changing vector currents.

$$[Q^{4+i5}, Q^{4-i5}] = Q^3 + \sqrt{3}\, Q^8 \qquad \left(\text{use } \begin{array}{l} f_{458} = \sqrt{3}/2 \\ f_{345} = 1/2 \end{array} \right)$$
$$= Q^{(e.m.)} + Y$$

Consider, for example, the leptonic decay of Σ^-.

$$\langle \Sigma^- | Q^3 | \Sigma^- \rangle = \langle \Sigma^- | [Q^{4+i5}, Q^{4-i5}] | \Sigma^- \rangle =$$
$$= - |\langle n | Q^{4+i5} | \Sigma^- \rangle|^2 + \underset{\substack{\uparrow \\ \Xi^- + \pi^-, \text{ etc.}}}{\sum_m |\langle m | Q^{4-i5} | \Sigma^- \rangle|^2} - \underset{\substack{\uparrow \\ n+\pi^0,\, p+\pi^-, \text{ etc.}}}{\sum_{m \neq n} |\langle m | Q^{4+i5} | \Sigma^- \rangle|^2}$$

If the summation terms could be ignored, then we would again have a nice simple relation:

$$- |\langle n | Q^{4+i5} | \Sigma^- \rangle|^2 = \langle \Sigma^- | Q^3 | \Sigma^- \rangle .$$

That these matrix elements are simply related by a C.G. coefficient would be expected from a naive application of the Cabibbo theory. In the exact SU(3) limit ($[H^{(sym.)}, Q^{4,5}] = 0$), n would be degenerate with Σ^-, and there would be no contribution from the summation terms. But in reality

$$[H, Q^{4,5}] = [H^{(sym)} + H^{(m.s.)}, Q^{4,5}] = [H^{(m.s.)}, Q^{4,5}].$$

where m.s. stands for "medium strong." So using

$$\langle m | Q^{4-i5} | \Sigma^- \rangle = \frac{\langle m | [H^{(m.s.)}, Q^{4-i5}] | \Sigma^- \rangle}{E_m - E_\Sigma}$$

we rewrite the summation as

$$-1 = \langle \Sigma^- | Q^3 | \Sigma^- \rangle = -|\langle n | Q^{4+i5} | \Sigma^- \rangle|^2$$

$$+ \sum_m \frac{|\langle m | [H^{(m.s.)}, Q^{4-i5}] | \Sigma^- \rangle|^2}{(E_m - E_\Sigma)^2}$$

$$- \sum_{m \neq n} \frac{|\langle m | [H^{(m.s.)}, Q^{4+i5}] | \Sigma^- \rangle|^2}{(E_m - E_\Sigma)^2}.$$

Let

$$\langle n | Q^{4+i5} | \Sigma^- \rangle_{ACTUAL} = r \langle n | Q^{4+i5} | \Sigma^- \rangle_{EXACT\,SU(3)} = -r.$$

In other words, r is a renormalization factor which measures the deviation from exact SU(3) symmetry. Observe now that $r^2 - 1 \approx 2\delta r$ is expressible in terms of

$$\sum_{m \neq n} \frac{|\langle m | [H^{(m.s.)}, Q^{4\pm i5}] | \Sigma^- \rangle|^2}{(E_m - E_\Sigma)^2}$$

and is <u>second</u> order in the medium strong interaction. In other words, $\delta r = 0$, $r = 1$ to first order in the SU(3) symmetry breaking interaction; i.e. there is no renormalization to first order in SU(3) breaking. To summarize, at zero momentum transfer the effect of SU(3) breaking on S-changing vector matrix elements is second order in the medium strong interaction. This is known as the Ademollo-Gatto theorem ('64), and the proof given here is due to Fubini and Furlan ('65).

I may mention that the Fubini-Furlan technique we have used to prove the Ademollo-Gatto theorem played a very important historical role. The famous Adler-Weisberger relation (which will be considered later in this course) was first derived by applying the Fubini-Furlan technique to the commutation relation

$$[Q_5^{1+i2}, Q_5^{1-i2}] = 2Q^3 .$$

CHAPTER III. VECTOR MESON UNIVERSALITY

HISTORY

Historically, predictions of vector mesons evolved along two different lines:

(1) Electromagnetic structure - Nambu ('57) postulated the existence of a neutral vector meson in order to explain the electromagnetic structure of nucleons. In the then conventional theory the nucleon structure was thought to be due mostly to pions. The virtual processes

$$p \leftrightarrow n + \pi^+$$
$$n \leftrightarrow p + \pi^-$$

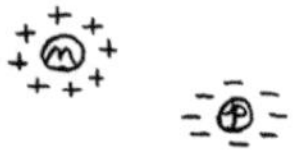

produce a positive pion cloud for p and a negative cloud for n. More quantitatively, if only pion pairs are important in the crossed channel (Look at the process "sideways".),

$$\gamma \leftrightarrow 2\pi \leftrightarrow \bar{N}N$$

the pion pair must be a $J^P = 1^-$, $T=1$, $T_3=0$ object coupled to the proton and the neutron with opposite sign ($T=1$, $T_3=0$, $\bar{N}N$ system is $\bar{p}p - \bar{n}n$). This gives $\langle r^2\rangle_p = -\langle r^2\rangle_n$, which is in serious disagreement with the experimental values.

$$\langle r^2\rangle_p \approx (0.7F)^2 , \quad \langle r^2\rangle_n \approx (0.0F)^2$$

So we need a contribution which has the same sign for both p and n in order to cancel the negative charge cloud of the n.

A $J^P = 1^-$, T=0 system (ω) satisfies this requirement since it couples to $\bar{p}p + \bar{n}n$.

This is how Nambu argued for the existence of a T=0, $J^{PG} = 1^{--}$ mesons.

The isovector analog of the above diagram is

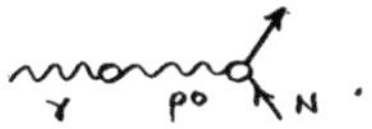

Indeed, Frazer and Fulco ('59) showed that the isovector part of the electromagnetic form factor, which is responsible for the difference between the p and n structures, is not well understood in terms of a theory based on uncorrelated pion pairs. So they were led to predict a T=1, $J^P = 1^-$ (p-wave) resonance in π-π scattering.

(2) <u>Universality and conserved currents</u> are important in electromagnetic and weak interactions, and it is natural to conjecture that these concepts also play important roles in strong interaction physics. On this basis were proposed (Sakurai '60) the strongly interacting (Y=0) vector mesons:

(i) T=1 coupled to isospin (like $j_\mu^{1,2,3}$)

(ii) T=0 " " hypercharge (" j_μ^8)

(iii) T=0 " " baryonic charge (" j_μ^0).

Which may appear as resonances in the invariant mass spectrum of 2π, 3π (and, if sufficiently massive, $K\bar{K}$) systems. It is natural to generalize to unitary symmetry by adding (Salam and Ward '61)

(iv) T= 1/2, Y=$\pm$1 strongly interacting vector mesons coupled to the strangeness changing currents (like $j_\mu^{4,5,6,7}$). Indeed, the first papers on the eightfold way (Gell-Mann, Ne'eman '61) were partially motivated by a desire to construct a higher symmetry scheme that accommodates vector mesons coupled to conserved and quasi-conserved currents. Shortly thereafter, mesons with the desired quantum numbers were discovered experimentally (1961-1963):

(i) T=1, Y=0, $J^{PG}=1^{-+}$; ρ(770)

(ii) T=0, Y=0, $J^{PG}=1^{--}$; ω(780), φ (1020)

(iii) T=1/2, Y= $\pm$1, $J^{P}=1^{-}$; K^* (890).

Gauge Principle and Universality

The original 1960 paper took advantage of the gauge principle proposed earlier by Yang and Mills ('54). Invariance under local isospin rotation

$$N \longrightarrow \left(1 + i\,\varepsilon_\alpha(x)\,\frac{\tau_\alpha}{2}\right) N$$

requires a T=1 vector meson coupled to the isospin. To see this we first note

$$-\bar N \gamma_\mu \partial_\mu N \longrightarrow -\bar N \gamma_\mu \partial_\mu N - i\,N \gamma_\mu \frac{\tau_\alpha}{2} N\,\partial_\mu \varepsilon_\alpha(x).$$

It is clear that the second term must somehow be cancelled if invariance is to be maintained.
This can be accomplished if there is a vector meson interaction of the form

$$i f_\rho\, \rho_\mu^\alpha\, \bar N \gamma_\mu \frac{\tau_\alpha}{2} N,$$

and the isospin rotation is accompanied by the gauge transformation,

$$\rho_\mu^\alpha \longrightarrow \rho_\mu^\alpha + \frac{1}{f_\rho}\partial_\mu \varepsilon_\alpha - \varepsilon_{\alpha\beta\gamma}\,\varepsilon_\beta\,\rho_\mu^\gamma$$

(where the last term is needed because ρ itself bears isospin).

The difficulty is, however, that the ρ has to be massless if the ρ free-field Lagrangian $(f_\rho = 0)$

$$\mathcal{L}^{(\rho)}_{free} = -\tfrac{1}{4}(\rho^\alpha_{\mu\nu})^2 - \tfrac{1}{2}m^2_\rho(\rho^\alpha_\mu)^2 \qquad [\rho^\alpha_{\mu\nu} = \partial_\mu\rho^\alpha_\nu - \partial_\nu\rho^\alpha_\mu]$$

is to be invariant under the above gauge transformation. Despite this difficulty we may argue that this is an interesting way to generate interactions, and so we repeat this for every isospin-bearing field. As in the case of the minimal electromagnetic coupling

$$\partial_\mu\psi^{(ch)} \longrightarrow \partial_\mu\psi^{(ch)} - ieA_\mu\psi^{(ch)}$$

we are then led to the ρ interaction by the substitution

$$\partial_\mu\psi \longrightarrow \partial_\mu\psi - if_\rho\vec{\rho}_\mu\cdot\vec{T}\psi,$$

where $\vec{T}$ is the isospin matrix acting on ψ (e.g. $\vec{\tau}/2$ for nucleons, $(T^\alpha)_{\beta\gamma} = -i\,\varepsilon_{\alpha\beta\gamma}$ for π and ρ). If you don't like the gauge principle, you can simply start with this "minimal" principle.

If N and π are taken as "fundamental", then

$$\mathcal{L}_{INT} = \vec{\rho}_\mu\cdot[i f_{\rho NN}\bar{N}\gamma_\mu\tfrac{\vec{\tau}}{2}N - f_{\rho\pi\pi}\vec{\pi}\times\partial_\mu\vec{\pi} - f_\rho\vec{\rho}_\nu\times\partial_\mu\vec{\rho}_\nu]$$

and "universality" requires $f_{\rho NN} = f_{\rho\pi\pi} = f_{\rho\rho\rho} = f_\rho$. Nowadays there are very many hadrons, and we don't know which ones (if any) are fundamental. Yet we may assume that the universality principle abstracted from gauge theory is true. In simple terms,

$$\frac{A\ (A \leftrightarrow A + \rho^0)}{A\ (B \leftrightarrow B + \rho^0)} = \frac{T_3^{(A)}}{T_3^{(B)}}$$

since the ρ^0 is supposed to be coupled to the 3rd component of isospin (T_3). This is analogous to the universality of electric charge.

$$\frac{A(e^+ \leftrightarrow e^+ + \gamma)}{A(p \leftrightarrow p + \gamma)} = \frac{Q(e^+)}{Q(p)} = 1$$

Actually this universality holds only at $q^2 = 0$. At large momentum transfer the p form factor is very different from that of the e^+ due to strong interactions. Likewise, ρ meson universality is expected to hold exactly at zero momentum transfer. Unfortunately, ρ is massive, and therefore $q^2 = 0$ cannot be realized for the emission of a real ρ. But if the vector meson form factor is slowly varying, the universality principle may still have some value even for the mass-shell coupling constants. More about this later.

Connection Between Vector Meson Dominance and Universality

The two seemingly unrelated lines of approach we have mentioned, one based on an attempt to understand the electromagnetic structure of hadrons - the other on universality and conserved currents, are actually intimately related. Gell-Mann and Zachariasen ('61) gave the key to this connection - vector meson dominance of the electromagnetic form factor.

Consider the e.m. form factor of the π^+ defined by

$$\langle p' | j_\mu^3(0) | p \rangle = \frac{1}{\sqrt{4\omega\omega'}} (p'+p)_\mu F_\pi(t)$$

(normalizing volume V=1 from now on; $t = -(p'-p)^2$). If $F_\pi(t) \to 0$ as $t \to \infty$, then we can write an unsubtracted dispersion relation

$$F_\pi(t) = \frac{1}{\pi} \int_{4m_\pi^2}^{\infty} dt' \frac{\operatorname{Im} F_\pi(t')}{t'-t}$$

with the constraint $F_\pi(0) = 1$ fixing the electric charge of the π^+ at zero momentum transfer. Suppose $F_\pi(t)$ is dominated by the ρ^0 meson, i.e., $\operatorname{Im} F_\pi(t') \propto \delta(t' - m_\rho^2)$, then

$$F_\pi(t) = \frac{m_\rho^2}{m_\rho^2 - t}.$$

Do the same for the proton isovector form factor

$$\langle p',s'|\, j^3_\mu(0)\,|p,s\rangle = \sqrt{\frac{m^2}{EE'}}\;\frac{1}{2}\Big(i\,\bar u'\gamma_\mu u\, F_1^{(V)}(t) + \underbrace{\text{anom.mom.term}}_{\text{zero at } q=0}\Big)$$

$$F_1^{(V)}(t) = \frac{m_\rho^2}{m_\rho^2 - t}$$

and compare diagrams.

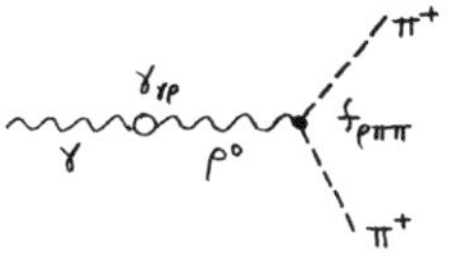

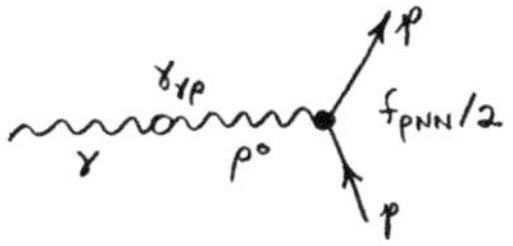

$$e\,F_\pi(t) = \frac{\gamma_{\gamma\rho}\, f_{\rho\pi\pi}}{m_\rho^2 - t} \qquad\qquad \frac{e}{2}\,F_1^{(V)}(t) = \frac{\gamma_{\gamma\rho}\, f_{\rho NN}/2}{m_\rho^2 - t}$$

(Only half of the electric charge of p is due to the isovector contribution.) At zero momentum transfer, $F_\pi(0) = F_1^{(V)}(0) = 1$, because the electric charges are universal. Therefore

$$\frac{\gamma_{\gamma\rho}\, f_{\rho\pi\pi}}{m_\rho^2} = \frac{\gamma_{\gamma\rho}\, f_{\rho NN}}{m_\rho^2} = e\,,$$

and we get

i) $f_{\rho\pi\pi} = f_{\rho NN} = \ldots$ (etc. for any particle with isospin)...$= f_\rho$

ii) $\gamma_{\gamma\rho} = e\, m_\rho^2/f_\rho$.

Thus universality is a consequence of complete ρ dominance of the isovector form factor. Moreover, we have learned that the effective coupling constant for $\rho^0 \leftrightarrow \gamma$ is given by $\gamma_{\gamma\rho} = e\,m_\rho^2/f_\rho$.

Since the electromagnetic interaction is given by

$$e\,A_\mu\Big(j^3_\mu + \frac{1}{\sqrt{3}}\, j^8_\mu\Big)$$

and $\langle 0|\, j^8_\mu\, |\rho^0\rangle = 0$ by isospin or G-parity, we have

$$\langle 0|\, j^3_\mu(x)\, |\rho^0\rangle = \frac{m_\rho^2}{f_\rho}\,\varepsilon_\mu\,\frac{e^{ip\cdot x}}{\sqrt{2\omega}}\,.$$

Recall that this matrix element is directly measurable in $\rho^0 \to e^+ + e^-$. So we may as well define the coupling constant f_ρ by the above matrix element relation regardless of whether complete ρ dominance holds.

Current-Field Identity (C.F.I.) for the ρ Meson

(Gell-Mann and Zachariasen '61; Kroll, Lee and Zumino '67).

As far as the matrix element between the vacuum and the ρ^0 state is concerned, j^3_μ acts just like the ρ^0 field, and it is tempting to set

$$j^\alpha_\mu = \frac{m_\rho^2}{f_\rho} \rho^\alpha_\mu .$$

This, the "current-field identity", conveniently summarizes the basic ingredients of vector meson universality and vector meson dominance as we'll see in a moment. But let us first demonstrate that this identity makes sense only if ρ is coupled to a conserved current. The ρ field Lagrangian

$$\mathcal{L}^{(\rho)} = -\frac{1}{4}(\partial_\mu \rho^\alpha_\nu - \partial_\nu \rho^\alpha_\mu)^2 - \frac{1}{2} m_\rho^2 (\rho^\alpha_\mu)^2 + \rho^\alpha_\mu J^{(\rho),\alpha}_\mu$$

gives the equation of motion

$$\partial_\nu(\partial_\nu \rho^\alpha_\mu - \partial_\mu \rho^\alpha_\nu) - m_\rho^2 \rho^\alpha_\mu = -J^{(\rho),\alpha}_\mu .$$

Taking the divergence, we get from C.F.I. and isospin conservation

$$\partial_\mu J^{(\rho),\alpha}_\mu = m_\rho^2 \partial_\mu \rho^\alpha_\mu = f_\rho \partial_\mu j^\alpha_\mu = 0 .$$

Therefore ρ is coupled to a conserved current. Also, since $\partial_\mu \rho^\alpha_\mu = 0$, the field equation becomes

$$(\Box - m_\rho^2)\rho^\alpha_\mu = -J^{(\rho),\alpha}_\mu .$$

This operator equation implies a corresponding matrix element relation:

$$-(p_B - p_A)^2 \langle B | \rho_\mu^\alpha | A \rangle - m_\rho^2 \langle B | \rho_\mu^\alpha | A \rangle = -\langle B | J_\mu^{(\rho),\alpha} | A \rangle$$

$$\langle B | \rho_\mu^\alpha | A \rangle = \frac{\langle B | J_\mu^{(\rho),\alpha} | A \rangle}{m_\rho^2 - t}$$

With C F I,

$$\langle B | j_\mu^\alpha | A \rangle = \frac{m_\rho^2}{f_\rho} \frac{\langle B | J_\mu^{(\rho),\alpha} | A \rangle}{m_\rho^2 - t}$$

and in particular $(\alpha = 3, t = 0)$

$$\langle B | j_\mu^3 | A \rangle_{t=0} = \frac{1}{f_\rho} \langle B | J_\mu^{(\rho),3} | A \rangle_{t=0} .$$

So at $t=0$ the source of the isovector photon field ($e j_\mu^3$) is identical to the source of the ρ^0— apart from a universal proportionality constant e/f_ρ. Moreover, at $t=0$, the isovector photon is known to be coupled universally to the 3rd component of the isospin. But since we have a strict proportionality between the source of the ρ^0 and the isovector current, we get

$$\langle A | J_0^{(\rho),3} | A \rangle |_{t=0} = f_\rho \langle A | j_0^3 | A \rangle |_{t=0} = f_\rho \langle A | Q^3 | A \rangle / V .$$

Thus the ρ^0 meson is also coupled universally to T_3 at $t=0$. This leads to

$$\left. \frac{A(A \leftrightarrow A + \rho^0)}{A(B \leftrightarrow B + \rho^0)} \right|_{t=0} = \frac{T_3^{(A)}}{T_3^{(B)}}$$

discussed earlier. Note that for the integrated ρ source we can write the operator equation $f_\rho Q^\alpha = \int d^3x \, J_0^{(\rho),\alpha}$.

It is interesting to note that our two important relations,

$$j_\mu^\alpha = \frac{m_\rho^2}{f_\rho} \rho_\mu^\alpha \qquad \text{(C F I)}$$

and

$$\langle B | j_\mu^\alpha | A \rangle = \frac{m_\rho^2}{m_\rho^2 - t} \frac{1}{f_\rho} \langle B | J_\mu^{(\rho),\alpha} | A \rangle ,$$

can be given a pictorial interpretation.

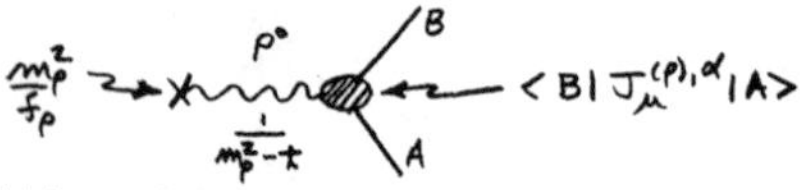

We'll come back to this point later.

Unfortunately, the coupling at t=o is not directly accessible, because of the finite mass of the ρ. In $\rho \to 2\pi$, for example, the coupling constant at $t = m_\rho^2$ is what is measured. To investigate these couplings for $t \neq o$, take out the uninteresting kinematical factors

$(e.g. \ (p+p')_\mu/\sqrt{4\omega\omega'}, \sqrt{\frac{m^2}{EE'}}, i\bar{u}\gamma_\mu u)$ common to both $\langle B|j_\mu^\alpha|A\rangle$ and $\langle B|J_\mu^{(\rho),\alpha}|A\rangle$. What remains is the "form factor" $F_{\rho AB}(t)$ is the vector meson form factor for $f_\rho^{-1}\langle B|J_\mu^{(\rho),\alpha}|A\rangle$. $F_{\gamma AB}^{(V)}(t)$ is the isovector e.m. form factor for $\langle B|j_\mu^\alpha|A\rangle$. (Throughout we ignore the complications arising from the instability of the ρ meson.) The two form factors are related by

$$F_{\gamma AB}^{(V)}(t) = \left(\frac{m_\rho^2}{m_\rho^2 - t}\right)\frac{F_{\rho AB}(t)}{F_{\rho AB}(0)}.$$

We note that for large $|t|$ the e.m. form factor goes to zero faster than the vector meson form factor by 1/t, e.g.:

If $F_{\rho AB}(t) \to$ const., then $F_{\gamma AB}^{(V)}(t) \to 0$ like $1/t$.

If " $1/t$, then " like $1/t^2$.

In this connection we may mention that the form factor measured in electron-proton scattering can be fitted by the formula

$$F_{\gamma NN}^{(V)}(t) = \left(\frac{m_\rho^2}{m_\rho^2 - t}\right)\left(\frac{\Lambda^2}{\Lambda^2 - t}\right)$$

where $\Lambda \approx$ 1 Bev (Massam and Zichichi '66).

Comparison with Dispersion-theoretic treatment (Gell-Mann and Zachariasen '61; Gell-Mann '62)

Complete ρ dominance in the sense of dispersion theory implies

$$F_{\gamma AB}^{(V)}(t) = \frac{m_\rho^2}{m_\rho^2 - t} .$$

This means $F_{\rho AB}(t)$ = const. = $F_{\rho AB}(0)$. From C F I we know that at t=o there is strict universality of the ρ coupling. If $F_{\rho AB}(t)$ varies little between t=o and t=m_ρ^2, then universal coupling is possible even at t= m_ρ^2.

Let us treat this problem in a more general manner. Taking the case A=B and considering just the "charge" coupling, we see that the ρAA coupling constant on the mass shell is given by

$$f_{\rho AA} = f_\rho \frac{F_{\rho AA}(m_\rho^2)}{F_{\rho AA}(0)} .$$

With the e.m. (charge) form factor written as

$$F_{\gamma AA}^{(V)}(t) = \frac{1}{\pi}\int \frac{\mathrm{Im}\,F(t')}{t'-t}\,dt'$$

separate out the residue at the isolated ρ pole by looking at the diagram for $\gamma_V \to \rho \to A\bar{A}$. We get

$$F_{\gamma AA}^{(V)}(t) = \left(\frac{m_\rho^2}{f_\rho}\right)\left(\frac{1}{m_\rho^2 - t}\right) f_{\rho AA} + \frac{1}{\pi}\int \frac{\mathrm{Im}\,F^{(\text{non-}\rho)}(t')}{t'-t}\,dt' .$$

Since the charge form factor for the isovector photon is unity at t=o, we have

$$F_{\gamma AA}^{(V)}(0) = 1 = \frac{f_{\rho AA}}{f_\rho} + \frac{1}{\pi}\int dt' \frac{\mathrm{Im}\,F^{(\text{non-}\rho)}(t')}{t'} .$$

Suppose the second term were negligible, then $f_{\rho AA} = f_\rho$. Repeat this argument for the charge form factor of the other hadrons. We get $f_{\rho AA} = f_{\rho BB} = f_{\rho CC} = \cdots = f_\rho$, and so we have universality even at $t = m_\rho^2$. To summarize, there are two equivalent ways of obtaining universality at $t = m_\rho^2$.

(i) In dispersion theory, assume that the non-ρ contributions are negligible.

(ii) In terms of C F I, assume slow variation of the vector meson form factor.

Im F (t) measures the amplitude for

$$\gamma^{(V)} \longrightarrow T=1, J^P=1^-, \text{ mass } = \sqrt{t} \text{ object} \longrightarrow \bar{A}A.$$

The hypothesis that $\text{Im}\, F^{(\text{non-}\rho)}_{(t)}$ is negligible can therefore be tested by performing a colliding beam experiment

$$e^+ + e^- \leftrightarrow \gamma \longrightarrow T=1, \ J^P=1^- \text{ object} \neq \rho$$

and checking whether anything but ρ's are produced copiously in the T= 1 channel. Experiments along this line are currently in progress in Siberia (Novosibirsk storage ring).

Renormalized vs. Unrenormalized ρ Field

[Gell-Mann, Zachariasen ('61); Kroll, Lee, Zumino ('67)]

C F I is a relation between the renormalized ρ field and j_μ^α whose 3rd component is the source of the isovector photon. But the source of the renormalized ρ field, $J_\mu^{(\rho),\alpha}$, is not proportional to j_μ^α (even though $\int d^3x\, J_0^{(\rho),\alpha} = f_\rho \int d^3x\, j_0^\alpha$). It is possible, however, to construct a field theory in which the source of the unrenormalized ρ field is proportional to j_μ^α.

Consider the Lagrangian of the unrenormalized ρ field ($\rho^{(0)}$).

$$\mathcal{L}^{(\rho^0)} = -\frac{1}{4}(\rho_{\mu\nu}^{(0)})^2 - \frac{1}{2}(m_\rho^{(0)}\rho_\mu^{(0)})^2 + f_\rho^{(0)} j_\mu \rho_\mu^{(0)} \quad \text{(suppress } \alpha=3)$$

Set $\rho_\mu^{(0)} = Z_3^{1/2}\rho_\mu$ (Z_3 = wave fcn. renorm. const.; ρ_μ= renormalized field) and rewrite $\mathcal{L}^{(\rho^0)}$ in terms of renormalized quantities.

$$\mathcal{L}^{(\rho^0)} = \mathcal{L}_{free}^{(\rho)} + \mathcal{L}_{INT}$$

$$\mathcal{L}_{free}^{(\rho)} = -\frac{1}{4}(\rho_{\mu\nu})^2 - \frac{1}{2}m_\rho^2\rho_\mu^2$$

$$\mathcal{L}_{INT} = -\frac{1}{4}(Z_3-1)(\rho_{\mu\nu})^2 - \frac{1}{2}(m_\rho^{(0)2}Z_3 - m_\rho^2)\rho_\mu^2 + f_\rho^{(0)} Z_3^{1/2} j_\mu \rho_\mu$$

This leads to the equation of motion,

$$\partial_\nu\rho_{\nu\mu} - m_\rho^2\rho_\mu = -(Z_3-1)\partial_\nu\rho_{\nu\mu} + (m_\rho^{(0)2}Z_3 - m_\rho^2)\rho_\mu - f_\rho^{(0)}Z_3^{1/2}j_\mu$$

$$= -J_\mu^{(\rho)} = \text{-source of renormalized } \rho;$$

therefore

$$-J_\mu^{(\rho)} = -(Z_3-1)(m_\rho^2\rho_\mu - J_\mu^{(\rho)}) + (m_\rho^{(0)2}Z_3 - m_\rho^2)\rho_\mu - f_\rho^{(0)}Z_3^{1/2}j_\mu$$

$$0 = \rho_\mu(-m_\rho^2 Z_3 + m_\rho^2 + m_\rho^{(0)2}Z_3 - m_\rho^2) + Z_3 J_\mu^{(\rho)} - f_\rho^{(0)}Z_3^{1/2}j_\mu$$

$$-J_\mu^{(\rho)} = (m_\rho^{(0)2} - m_\rho^2)\rho_\mu - f_\rho^{(0)}Z_3^{-1/2}j_\mu.$$

With $\partial_\mu\rho_\mu^{(0)} = \partial_\mu\rho_\mu = 0$ we have

$$\text{(i)}\quad (\Box - m_\rho^{(0)2})\rho_\mu^{(0)} = -f_\rho^{(0)}j_\mu$$

$$\text{(ii)}\quad (\Box - m_\rho^2)\rho_\mu = -J_\mu^{(\rho)} = -f_\rho^{(0)}Z_3^{-1/2}j_\mu + (m_\rho^{(0)2} - m_\rho^2)\rho_\mu.$$

From (ii)

$$\langle B|\rho_\mu|A\rangle = \frac{f_\rho^{(0)}Z_3^{-1/2}\langle B|j_\mu|A\rangle}{m_\rho^{(0)2} - t}.$$

But we also have

$$\langle B|\rho_\mu|A\rangle = \frac{\langle B|J_\mu^{(\rho)}|A\rangle}{m_\rho^2 - t}.$$

So

$$\langle B | j_\mu | A \rangle = \left(\frac{m_\rho^{(0)2} - t}{m_\rho^2 - t} \right) \frac{Z_3^{1/2}}{f_\rho^{(0)}} \langle B | J_\mu^{(\rho)} | A \rangle .$$

In general, using the spectral representation for the ρ meson propagator, we can show $m_\rho^{(0)} > m_\rho$. For a divergent theory, $m_\rho^{(0)} = \infty$, and the above formula becomes the familiar relation

$$\langle B | j_\mu | A \rangle = \frac{m_\rho^2 / f_\rho}{m_\rho^2 - t} \langle B | J_\mu^{(\rho)} | A \rangle$$

where

$$\frac{m_\rho^2}{f_\rho} = Z_3^{1/2} \frac{m_\rho^{(0)2}}{f_\rho^{(0)}} .$$

Thus C F I is derivable from a field theory in which the unrenormalized ρ field is coupled to the isovector electromagnetic current (source of $\rho^{(0)}$ is proportional to j_μ) and the bare mass $m_\rho^{(0)} = \infty$. If $m_\rho^{(0)}$ were finite, $F_{\gamma AB}^{(V)}$ (t) would have a "universal zero" at t= $m_\rho^{(0)2}$.

$$F_{\gamma AB}^{(V)}(t) = \left(\frac{m_\rho^{(0)2} - t}{m_\rho^2 - t} \right) \left(\frac{m_\rho}{m_\rho^{(0)}} \right)^2 \frac{F_{\rho AB}(t)}{F_{\rho AB}(0)}$$

$$= \left(\frac{m_\rho^2}{m_\rho^2 - t} \right) \left(\frac{F_{\rho AB}(t)}{F_{\rho AB}(0)} \right) \left(1 - \frac{t}{m_\rho^{(0)2}} \right)$$

It was G. F. Chew who first argued that "elementary particle" theory can be distinguished from "dynamical" theory just by looking for a zero in the form factor. Gell-Mann and Zachariasen have subsequently shown that to convert elementary particle theory into dynamical theory, $m_\rho^{(0)} \to \infty$ is all that is needed. It is amusing that if $m_\rho^{(0)}$ were finite, we could "measure" it by looking for a zero in the isovector electromagnetic form factor for $t > 0$. Taking $m_\rho^{(0)} \to \infty$ gives us the usual "dynamical" theory with smooth behavior.

Determination of the ρ Meson Coupling Constant (Sakurai '66)

The most practical way of testing ρ meson universality is simply to measure $f_{\rho AA}$ for various processes. If the variation of $F_{\rho AA}(t)$ between $t=0$ and $t=m_\rho^2$ is negligible, then we must have $f_{\rho AA} = f_{\rho BB} = \ldots\ldots = f_\rho$. So consider the following determinations of the ρ coupling constants:

(A) $f_{\rho\pi\pi}$ is directly measurable from $\rho \to \pi\pi$.

$$\Gamma(\rho \to \pi\pi) = \frac{2}{3}\frac{f_{\rho\pi\pi}^2}{4\pi}\frac{p_{\pi\pi}^3}{m_\rho^2}$$

$$\Gamma_\rho = (128 \pm 5)\,\mathrm{Mev}\ ,\quad m_\rho = (774 \pm 3)\,\mathrm{Mev}$$

$$\frac{f_{\rho\pi\pi}^2}{4\pi} = 2.5 \pm 0.1$$

(B) The ρ exchange contribution to π-N scattering measures $f_{\rho\pi\pi} f_{\rho NN}$.

(i) Complete ρ dominance (at threshold) of the S-wave scattering leads to (Sakurai '60)

$$\underset{T=1/2}{a_1} - \underset{T=3/2}{a_3} = 3\,\frac{f_{\rho\pi\pi} f_{\rho NN}}{4\pi}\,\frac{m_\pi m_N}{m_\pi + m_N}\,\frac{1}{m_\rho^2}\ .$$

(This assumption may sound a little naive, but later we'll justify this to some extent using current algebra.)

$$a_1 - a_3 = (.271 \pm .007)/m_\pi$$

$$\frac{f_{\rho\pi\pi} f_{\rho NN}}{4\pi} = 2.8 \pm 0.1$$

(ii) Using partial wave dispersion relations, estimate the ρ contribution from the <u>energy dependence</u> of the S-wave scattering amplitudes.

$$\frac{f_{\rho\pi\pi} f_{\rho NN}}{4\pi} = 2.85 \pm 0.3$$

(Hamilton. '67. See also Bowcock, Cottingham and Lurié '61)

(C) N-N scattering measures $f_{\rho NN}$. But the determination is rather poor because it is difficult to isolate the ρ exchange from other contributions.

$$\frac{f^2_{\rho NN}}{4\pi} = 2.1 - 5.1$$ (Bryan and Scott '64; Scotti and Wong '66.)

(D) The Gell-Mann-Sharp-Wagner ('62) model for ω decay.

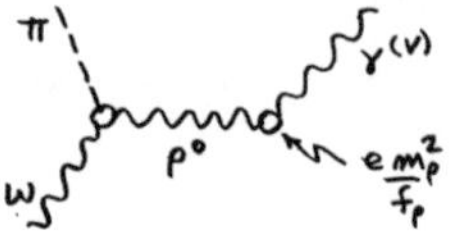

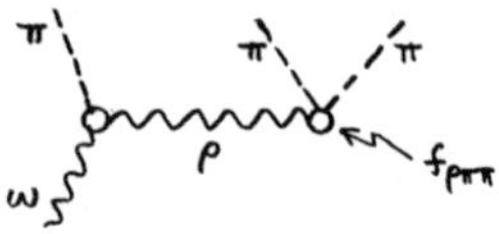

Note that the $\omega\rho\pi$ coupling constant is common to both diagrams.

$$\frac{\Gamma(\omega \to \pi + \gamma^{(V)})}{\Gamma(\omega \to 3\pi)} = \frac{e^2/4\pi}{\frac{f^2_{\rho\pi\pi}}{4\pi}\,\frac{f^2_\rho}{4\pi}} \times \underbrace{(98 \pm 6)}$$

phase space and other kinematical factors

experimental ratio = $\dfrac{(9 \pm 1)\%}{90\%} \Rightarrow \dfrac{f_\rho f_{\rho\pi\pi}}{4\pi} = 2.7 \pm 0.2.$

(E) The leptonic pair decay of ρ^0 measures $e\,m^2_\rho/f_\rho$ which can be taken as a definition of f_ρ (Nambu and Sakurai '62).

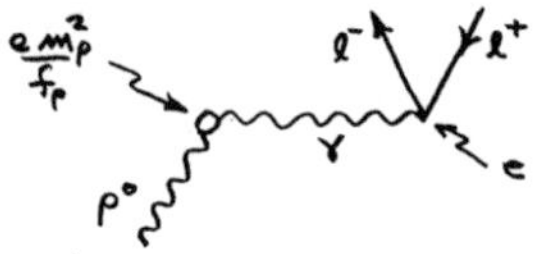

$$\Gamma(\rho^\circ \to \ell^+\ell^-) = \frac{(e^2/4\pi)^2}{f_\rho^2/4\pi}\left(\frac{m_\rho}{3}\right)\underbrace{\left[1+2\left(\frac{m_\ell}{m_\rho}\right)^2\right]\left[1-4\left(\frac{m_\ell}{m_\rho}\right)^2\right]^{1/2}}_{\approx 1},$$

where $\ell^+ \ell^-$ may stand for $\mu^+\mu^-$ or $e^+ e^-$.

$$\frac{\Gamma(\rho \to \mu^+\mu^-)}{\Gamma(\rho \to \pi^+\pi^-)} = \frac{e^4}{f_\rho^2 f_{\rho\pi\pi}^2}\left(\frac{m_\rho^2}{2\, p_{\pi\pi}^3}\right)$$

experimental ratio = $(5.1 \pm 1.2) \times 10^{-5}$

$$\frac{f_\rho f_{\rho\pi\pi}}{4\pi} = 2.3 \pm 0.3$$

All the above determinations are consistent with $f_{\rho\pi\pi} = f_{\rho NN} = f_\rho$. Complete ρ dominance at all values of t, however, is not consistent with the e.m. form factor deduced from $e-p$ scattering data because the observed form factor goes to zero faster than $1/t$ as $|t| \to \infty$. Perhaps ρ dominance is alright for $t = 0$ — m_ρ^2 but not good for large negative t. More recently, the π form factor has been estimated from $e^- + p \longrightarrow e^- + M + \pi^+$. Extracting the one pion exchange contribution,

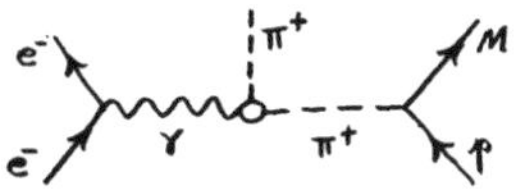

a Cornell group has obtained

$$r_\pi = \sqrt{6\left(\frac{dF}{dt}\right)_{t=0}} = 0.8 \pm 0.1 \text{ Fermi;}$$

whereas complete ρ dominance implies $r_\pi = \sqrt{6}/m_\rho \approx .6F$. In any case it is desirable to have more accurate and reliable data on $\Gamma(\rho \to \ell^+\ell^-)$ and $\Gamma(\rho \to \pi\pi)$. The best way to measure these is $e^+ + e^- \to \rho \to \pi\pi$, where there is no background problem due to other final state particles.

Once determined, f_ρ can be used in many other places. Consider, for example, the high-energy photoproduction of ρ°.

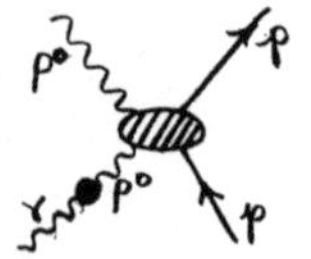

Both the differential cross section and the ρ spin alignment are consistent with this picture ("diffraction dissociation" mechanism proposed by Ross, Stodolsky, Freund '66) We expect

$$\frac{\sigma(\gamma p \to \rho^\circ p)}{\sigma(\rho^\circ p \to \rho^\circ p)} = \frac{e^2}{f_\rho^2} = \frac{1/137}{f_\rho^2/4\pi} .$$

Presumably diffraction scattering is dominant in $\rho^\circ p \to \rho^\circ p$ at high energies. So from

$$\frac{d\sigma}{dt} = \left(\frac{d\sigma}{dt}\right)_{t=0} e^{at}$$

deduce

$$\sigma_{tot}^2(\rho^\circ p) = 16\pi a\, \sigma_{eL}(\rho^\circ p) .$$

With this we can test the quark model prediction (Lipkin, etc.)

$$\sigma_{tot}(\rho^\circ p) = \frac{1}{2}\left[\sigma_{tot}(\pi^- p) + \sigma_{tot}(\pi^+ p)\right].$$

Conversely, guessing $\sigma_{tot}(\rho^\circ p)$ from the quark model and measuring a from the observed slope of the photoproduction cross section, we get (Joos '67)

$$\frac{f_\rho^2}{4\pi} = 2.2 \pm 0.2 .$$

But this should not be taken too seriously. (More recently, by studying the A dependence of γ+ Nucleus $\longrightarrow \rho^\circ$+ Nucleus, $\sigma_{tot}(\rho N)$ has been deduced by Drell and Trefil '66. The most recent data by a DESY group gives $\sigma_{tot}(\rho N) = (31\pm 2)$ mb in remarkable agreement with the quark model estimate.)

Gauge Invariance and the em_ρ^2/f_ρ Prescription

The prescription of inserting em_ρ^2/f_ρ at a γ-ρ^0 junction suggests the effective interaction

$$\mathcal{L}_{eff} = \frac{em_\rho^2}{f_\rho} \rho_\mu^3 A_\mu .$$

But is this consistent with gauge invariance? The gauge transformation $A_\mu \rightarrow A_\mu + \partial_\mu \Lambda$ gives an extra term

$$\delta \mathcal{L} = \frac{em_\rho^2}{f_\rho} \rho_\mu \partial_\mu \Lambda$$

(Omit isospin index, neutral ρ implied.). One may argue that this does not violate gauge invariance because $\partial_\mu \rho_\mu = 0$. But this argument is <u>false</u>. $\partial_\mu \rho_\mu = 0$ is an identity that follows by virtue of the equation of motion when $\partial_\mu J_\mu^{(\rho)} = 0$. One cannot use the equation of motion in writing down the Lagrangian. As a classic example of this, $\mathcal{L}_N = -\bar{N}(\gamma_\mu \partial_\mu + m)N$ would be zero by the Dirac equation.

To show that there is real trouble with the em_ρ^2/f_ρ prescription, we compute the modification of the photon propagator due to the $\rho_\mu A_\mu$ interaction.

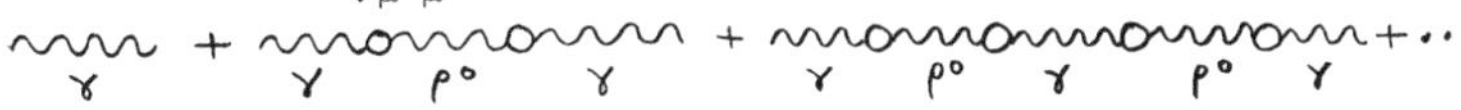

For small k it is all right to ignore the $k_\mu k_\nu$ term in the ρ propagator. The $\delta_{\mu\nu}$ part of the photon propagator is

$$D(k^2) = \frac{1}{k^2} + \frac{1}{k^2}\left(\frac{e m_\rho^2}{f_\rho}\right)^2 \frac{1}{k^2+m_\rho^2}\,\frac{1}{k^2} +$$

$$+ \frac{1}{k^2}\left(\frac{e m_\rho^2}{f_\rho}\right)^2 \frac{1}{k^2+m_\rho^2}\,\frac{1}{k^2}\left(\frac{e m_\rho^2}{f_\rho}\right)^2 \frac{1}{k^2+m_\rho^2}\,\frac{1}{k^2} + \cdots .$$

Sum this series using the formula,

$$\frac{1}{A-B} = \frac{1}{A} + \frac{1}{A}B\frac{1}{A} + \frac{1}{A}B\frac{1}{A}B\frac{1}{A} + \cdots .$$

Thus

$$D(k^2) = \frac{1}{k^2 - \left(\frac{e m_\rho^2}{f_\rho}\right)^2 \frac{1}{k^2+m_\rho^2}} \approx \frac{1}{k^2 - \left(\frac{e}{f_\rho}\right)^2 m_\rho^2}$$

for small k. This should be compared with the propagator of a massive particle which reads in my metric

$$D(k^2) = \frac{1}{k^2+m^2} .$$

That the photon has mass is bad enough -- but an imaginary mass!

$$m^2_{\text{PHOTON}} = -\left(\frac{e}{f_\rho}\right)^2 m_\rho^2 \quad ?$$

For this reason the em_ρ^2/f_ρ prescription was severely criticized by Feldman and Matthews ('63). Yet people continued using the prescription with the pet phrase, "In the dispersion-theoretic formalism it is all right ...".

Recently Kroll, Lee and Zumino ('67) succeeded in constructing a Lagrangian formalism in which $(em_\rho^2/f_\rho)\,\rho_\mu^3$ is the source of the isovector part of the Maxwell field. Nobody objects to the $F_{\mu\nu}\,\rho_{\mu\nu}$ interaction, which is manifestly gauge invariant. But this interaction alone will not do the job because it vanishes for real photons with $q^2 = 0$. So we add another interaction $(e/f_\rho)\, A_\mu\, J_\mu^{(\rho)}$. It is easy to see that if the ρ interaction is generated by the minimal principle

$$\partial_\mu \psi \longrightarrow \partial_\mu \psi - i f_\rho \vec{\rho}_\mu \cdot \vec{T} \psi ,$$

then this interaction is also gauge invariant; for instance, the nucleon part of this interaction is

$$\frac{e}{f_\rho} A_\mu i f_\rho \bar{N} \gamma_\mu \frac{\tau_3}{2} N = i \frac{e}{2} A_\mu \bar{N} \gamma_\mu \tau_3 N ,$$

which is just the isovector part of the "old-fashioned" electromagnetic interaction of the nucleon, which is, of course, gauge invariant. There is a particular linear combination of the $F_{\mu\nu}\rho_{\mu\nu}$ interaction and the $A_\mu J_\mu^{(\rho)}$ interaction that gives the desired result. The relevant part of the Lagrangian is taken to be

$$\mathcal{L} = -\frac{1}{4}(F_{\mu\nu})^2 - \frac{1}{4}(\rho_{\mu\nu})^2 - \frac{1}{2}(m_\rho \rho_\mu)^2 + \rho_\mu J_\mu^{(\rho)}$$
$$+ \frac{e}{f_\rho}\left[-\frac{1}{2} F_{\mu\nu}\rho_{\mu\nu} + A_\mu J_\mu^{(\rho)}\right].$$

Note that the terms of order e are just what we get by the substitution

$$\rho_\mu \longrightarrow \rho_\mu + \frac{e}{f_\rho} A_\mu$$

in the $(\rho_{\mu\nu})^2$ and the $\rho_\mu J_\mu^{(\rho)}$ part [but not in the $(\rho_\mu)^2$ part] of the ρ meson Lagrangian. From this Lagrangian we get the equation of motion

$$\partial_\nu F_{\nu\mu} = \frac{\delta}{\delta A_\mu}\left\{ \frac{e}{f_\rho}\left[-\frac{1}{2}F_{\mu\nu}\rho_{\mu\nu} + A_\mu J_\mu^{(\rho)}\right]\right\}$$

$$= \frac{e}{f_\rho}\left(\partial_\nu \rho_{\nu\mu} + J_\mu^{(\rho)}\right)$$

$$= \frac{e m_\rho^2}{f_\rho}\rho_\mu ,$$

where we have used

$$-\frac{1}{2}\rho_{\mu\nu}(\partial_\mu A_\nu - \partial_\nu A_\mu) = -\rho_{\nu\mu}\partial_\nu A_\mu \rightarrow \partial_\nu \rho_{\nu\mu} A_\mu$$

and the equation of motion for the ρ field. Note that the source of the Maxwell field is $(em_\rho^2/f_\rho)\rho_\mu$, just as advertized.

It is also possible to rewrite this Lagrangian in a more convenient form:

$$\mathcal{L} = -\frac{1}{4}(F'_{\mu\nu})^2 - \frac{1}{4}(\rho'_{\mu\nu})^2 - \frac{1}{2}(m_\rho \rho'_\mu)^2 + \rho'_\mu J_\mu^{(\rho)}$$
$$+ \left(\frac{e' m_\rho^2}{f_\rho}\right)\rho'_\mu A'_\mu - \frac{1}{2}\left(\frac{e'}{f_\rho}\right)^2 m_\rho^2 A'^2_\mu$$

$$\rho'_\mu = \rho_\mu - \frac{e}{f_\rho}A_\mu \qquad\qquad A'_\mu = \sqrt{1-\left(\frac{e}{f_\rho}\right)^2}\,A_\mu$$

$$e' = e\sqrt{1-\left(\frac{e}{f_\rho}\right)^2}.$$

The $\rho \cdot A$ term by itself violates gauge invariance, but is all right when combined with the "photon mass term,"

$$-\frac{1}{2}\left(\frac{e'}{f_\rho}\right)^2 m_\rho^2 A_\mu^2 .$$

To illustrate this, we now recalculate the photon propagator using the new interaction Lagrangian that includes the photon mass term.

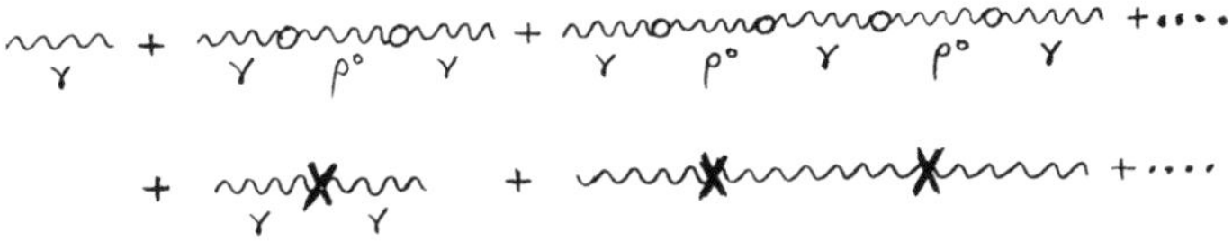

$$D(k^2) = \frac{1}{k^2} + \frac{1}{k^2}\left(\frac{e' m_\rho^2}{f_\rho}\right)^2 \frac{1}{k^2 + m_\rho^2}\,\frac{1}{k^2} + \cdots$$

$$- \frac{1}{k^2}\left(\frac{e'}{f_\rho}\right)^2 m_\rho^2 \,\frac{1}{k^2} + \cdots$$

$$= \frac{1}{k^2} + \frac{1}{k^2}\left(\frac{e' m_\rho}{f_\rho}\right)^2 \left(\frac{-k^2}{k^2 + m_\rho^2}\right)\frac{1}{k^2} + \cdots$$

$$= \frac{1}{k^2 + \left(\frac{e'}{f_\rho}\right)^2 m_\rho^2 \left(\frac{k^2}{k^2 + m_\rho^2}\right)} \approx \left[1 - \left(\frac{e}{f_\rho}\right)^2\right]\frac{1}{k^2}$$

for small k.

The photon mass contribution cancels exactly with the negative $(\text{mass})^2$ term due to the $\rho \cdot A$ interaction, and we are simply left with a change in the e.m. coupling constant.

$$e^2 \rightarrow e^2\left[1 - \left(\frac{e}{f_\rho}\right)^2\right]$$

The T = 1 hadronic contribution to charge renormalization is completely finite,

$$\frac{\delta e^2}{e^2} = \left(\frac{e}{f_\rho}\right)^2 = 0.28\% .$$

Note that only $\approx$ 0.3% of 1/137 is due to the strong interaction. This discussion ignores the isoscalar contribution, but using SU(3) and ω-φ mixing, we can show that the isoscalar contribution is even smaller.

Historically, Kroll, Lee and Zumino ('67) first obtained the inequality,

$$\frac{\delta e^2}{e^2} < \left(\frac{e\, m_\rho}{f_\rho 2 m_\pi}\right)^2 \approx 2.4\% .$$

to point out the finiteness of the hadronic contribution to charge renormalization without committing themselves to complete ρ dominance, as we have done. Their limit is extremely conservative since this value is obtained when one assumes that the T=1, $J^P = 1^-$ spectral function is concentrated at the 2π threshold - rather than at ρ.

We want to conclude this section by commenting on the "Feynman graph" interpretation of the γ-ρ junction. We have already seen that there are two equivalent ways of writing down the gauge-invariant electromagnetic couplings of hadrons compatible with CFI:

(1) $\left(\frac{e}{f_\rho}\right)\left(-\frac{1}{2}F_{\mu\nu}\rho_{\mu\nu} + J_\mu^{(\rho)}A_\mu\right)$

(ii) $\left(\frac{e m_\rho^2}{f_\rho}\right)\rho_\mu A_\mu - \frac{1}{2}\left(\frac{e}{f_\rho}\right)^2 m_\rho^2 A_\mu^2 .$

If we follow the Feynman rule with the interaction (i) we get for $\gamma + A \to B$

$$\left[-\left(\frac{e}{f_\rho}\right)\frac{q^2}{q^2+m_\rho^2} + \frac{e}{f_\rho}\right]\langle B| J_\mu^{(\rho)} |A\rangle .$$

Note that the first term due to the direct interaction vanishes at $q^2 = 0$, and it is therefore very important to take into account the effect of the "contact" $J_\mu^{(\rho)} A_\mu$ interaction. If, on the other hand, we prefer to work with the interaction (ii), to order e we need to consider only the $\rho_\mu A_\mu$ interaction; this time the direct γ-ρ interaction does not vanish at $q^2 = 0$, and the result is just

$$\left(\frac{e}{f_\rho}\right)\frac{m_\rho^2}{q^2+m_\rho^2}\langle B| J_\mu^{(\rho)} |A\rangle .$$

The two approaches are obviously equivalent because of the identity

$$-\frac{q^2}{q^2+m_\rho^2} + 1 \equiv \frac{m_\rho^2}{q^2+m_\rho^2} .$$

Gauge Field Algebra

In the last chapter we discussed the commutation relations among j_μ^α and $j_{5\mu}^\alpha$ using the free-field quark model as a guide. The usefulness of CFI, $j_\mu^\alpha = \frac{m_\rho^2}{f_\rho}\rho_\mu^\alpha$ (with $\alpha = 1,2,3$), naturally leads to the question: Can the same (or similar) commutation relations be derived from the "canonical" field theory of the ρ meson field? An attempt along this line was first made by Lee, Zumino and Weinberg ('67), whose treatment we now follow.

The basic Lagrangian we start with is the one obtained by applying the "minimal" substitution

$$\partial_\mu \psi \rightarrow \partial_\mu \psi - i f_\rho \rho_\mu^\alpha T_\alpha \psi$$

$$\partial_\mu \rho_\nu^\alpha \rightarrow \partial_\mu \rho_\nu^\alpha + f_\rho \varepsilon_{\alpha\beta\gamma} \rho_\mu^\beta \rho_\nu^\gamma$$

to the free Lagrangian of the ρ field and the "matter" fields (i.e. the basic fields other than the ρ field).

$$\mathcal{L} = -\frac{1}{4} \rho_{\mu\nu}^\alpha \rho_{\mu\nu}^\alpha - \frac{1}{2} m_\rho^2 \rho_\nu^\alpha \rho_\nu^\alpha + \mathcal{L}_m$$

where $\rho_{\mu\nu}$ now stands for

$$\rho_{\mu\nu}^\alpha = \partial_\mu \rho_\nu^\alpha - \partial_\nu \rho_\mu^\alpha + f_\rho \varepsilon_{\alpha\beta\gamma} \rho_\mu^\beta \rho_\nu^\gamma$$

(Note the last term.) It is understood that the coupling of the ρ field to the matter fields is contained in $\mathcal{L}_m$ while the mutual interactions of the ρ fields (à la Yang and Mills) are hidden in the $\rho_{\mu\nu}\rho_{\mu\nu}$ term. We first evaluate

$$\frac{\partial \mathcal{L}}{\partial(\partial_\mu \rho_\nu^\alpha)} = -\frac{1}{2}(\rho_{\mu\nu}^\alpha - \rho_{\nu\mu}^\alpha) = -\rho_{\mu\nu}^\alpha$$

and

$$\frac{\partial \mathcal{L}}{\partial \rho_\nu^\alpha} = -f_\rho \varepsilon_{\alpha\beta\gamma} \rho_{\mu\nu}^\beta \rho_\mu^\gamma - m_\rho^2 \rho_\nu^\alpha + \frac{\partial \mathcal{L}_m}{\partial \rho_\nu^\alpha} ;$$

where we have assumed that $\mathcal{L}_m$ does not contain the derivatives of the ρ field, as is the case in the minimal coupling theory. The equation of motion for the ρ field can be obtained immediately by writing down the Euler-Lagrange equation

$$\frac{\partial}{\partial x_\mu}\left(\frac{\partial \mathcal{L}}{\partial(\partial_\mu \rho_\nu^\alpha)}\right) = \frac{\partial \mathcal{L}}{\partial \rho_\nu^\alpha} .$$

We get

$$m_\rho^2 \rho_\nu^\alpha = \partial_\mu \rho_{\mu\nu}^\alpha - f_\rho \varepsilon_{\alpha\beta\gamma} \rho_{\mu\nu}^\beta \rho_\mu^\gamma + \frac{\partial \mathcal{L}_m}{\partial \rho_\nu^\alpha} .$$

In the canonical field theory we must obtain the momentum variable (analogous to $p = \partial L/\partial \dot{q}$) conjugate to the field variable (analogous to q). This can be done by setting μ=4 in our earlier expression for $\partial\mathcal{L}/\partial(\partial_\mu \rho_\nu^\alpha)$. Denoting the canonical momentum conjugate to ρ_k^α by Π_k^α, we get

$$\Pi_k^\alpha = -i \frac{\partial \mathcal{L}}{\partial\left(\frac{\partial \rho_k^\alpha}{\partial x_4}\right)} = -i \rho_{k4}^\alpha .$$

We can rewrite the field equation with ν =4 as follows

$$m_\rho^2 \rho_4^\alpha = i \partial_k \Pi_k^\alpha - i f_\rho \varepsilon_{\alpha\beta\gamma} \Pi_k^\beta \rho_k^\gamma + \frac{\partial \mathcal{L}_m}{\partial \rho_4^\alpha} .$$

It is crucial to note here that the fourth component of the field is expressible completely in terms of the space components ρ_k^α and their canonical momenta Π_k^α.

In the canonical field theory the space components of the ρ field are taken to be independent field variables; hence they commute at equal times

$$[\rho_k^\alpha(x), \rho_\ell^\beta(x')]_{x_0=x_0'} = 0 .$$

Similarly the canonical momenta Π_k^α's are assumed to satisfy

$$[\Pi_k^\alpha(x), \Pi_\ell^\beta(x')]_{x_0=x_0'} = 0 .$$

The basic nontrivial commutation relations (analogous to $[q, p] = i$) are

$$[\rho_k^\alpha(x), \pi_\ell^\beta(x')]_{x_0=x_0'} = i\,\delta_{\alpha\beta}\,\delta_{k\ell}\,\delta^{(3)}(\vec{x}-\vec{x}').$$

Armed with these relations, we can evaluate the commutator between ρ_4^α and ρ_ℓ^β.

$$m_\rho^2[\rho_4^\alpha(x), \rho_\ell^\beta(x')]_{x_0=x_0'} = i\partial_k[\pi_k^\alpha(x), \rho_\ell^\beta(x')] - i f_\rho \varepsilon_{\alpha\delta\gamma}[\pi_k^\delta(x), \rho_\ell^\beta(x')]\rho_k^\gamma(x)$$

$$= -f_\rho\,\varepsilon_{\alpha\beta\gamma}\,\rho_\ell^\gamma(x)\,\delta^{(3)}(\vec{x}-\vec{x}') + \delta_{\alpha\beta}\,\partial_\ell\,\delta^{(3)}(\vec{x}-\vec{x}')$$

It is somewhat more involved to derive the commutator between time components. We first note that if the ρ interaction is to be generated by the minimal principle, $\partial\mathcal{L}_m/\partial\rho_\mu^\alpha$ takes the form

$$\frac{\partial\mathcal{L}_m}{\partial\rho_\mu^\alpha} = i f_\rho \bar{\psi}\gamma_\mu T_\alpha\psi\,,$$

hence

$$\frac{\partial\mathcal{L}_m}{\partial\rho_4^\alpha} = i f_\rho \psi^\dagger T_\alpha \psi$$

where we have assumed (for illustrative purposes) that the basic matter field is a spin 1/2 field. We can now evaluate the desired commutator as follows:

$$m_\rho^4[\rho_4^\alpha(x), \rho_4^\beta(x')]_{x_0=x_0'} =$$

$$= [\,i\partial_k\pi_k^\alpha(x) - i f_\rho\varepsilon_{\alpha\gamma\delta}\pi_k^\gamma(x)\rho_k^\delta(x) + i f_\rho\psi^\dagger(x)T_\alpha\psi(x)\,,$$
$$i\partial_\ell'\pi_\ell^\beta(x') - i f_\rho\varepsilon_{\beta\epsilon\eta}\pi_\ell^\epsilon(x')\rho_\ell^\eta(x') + i f_\rho\psi^\dagger(x')T_\beta\psi(x')]_{x_0=x_0'}$$

$$= f_\rho \varepsilon_{\beta\epsilon\eta} [\partial_k \Pi_k^\alpha(x), \rho_\ell^\eta(x')] \Pi_\ell^\epsilon(x') + f_\rho \varepsilon_{\alpha\gamma\delta} \Pi_k^\gamma(x) [\rho_k^\delta(x), \partial'_\ell \Pi_\ell^\beta(x')]$$

$$- f_\rho^2 \varepsilon_{\alpha\gamma\delta} \varepsilon_{\beta\epsilon\eta} \Pi_k^\gamma(x) [\rho_k^\delta(x), \Pi_\ell^\epsilon(x')] \rho_\ell^\eta(x') - f_\rho^2 \varepsilon_{\alpha\gamma\delta} \varepsilon_{\beta\epsilon\eta} \Pi_\ell^\epsilon(x') [\Pi_k^\gamma(x), \rho_\ell(x')] \rho_k(x)$$

$$- f_\rho^2 [\psi^+(x) T_\alpha \psi(x), \psi^+(x') T_\beta \psi(x')]$$

$$= -i f_\rho \varepsilon_{\beta\epsilon\alpha} \partial_k [\delta^{(3)}(\vec{x}-\vec{x}')] \Pi_k^\epsilon(x') + i f_\rho \epsilon_{\alpha\gamma\beta} \partial'_k [\delta^{(3)}(\vec{x}-\vec{x}')] \Pi_k^\gamma(x)$$

$$- i f_\rho^2 \varepsilon_{\alpha\gamma\delta} \varepsilon_{\beta\delta\eta} \Pi_k^\gamma(x) \rho_k^\eta(x) \delta^{(3)}(\vec{x}-\vec{x}') + i f_\rho^2 \varepsilon_{\alpha\gamma\delta} \varepsilon_{\beta\epsilon\gamma} \Pi_k^\epsilon(x) \rho_k(x) \delta^{(3)}(\vec{x}-\vec{x}')$$

$$- i f_\rho^2 \varepsilon_{\alpha\beta\gamma} \psi^+(x) T_\gamma \psi(x) \delta^{(3)}(\vec{x}-\vec{x}')$$

$$= \left[-i f_\rho \varepsilon_{\alpha\beta\gamma} \partial_k \Pi_k^\gamma(x) + i f_\rho^2 \varepsilon_{\alpha\beta\gamma} \varepsilon_{\gamma\delta\epsilon} \Pi_k^\delta(x) \rho_k^\epsilon(x) - i f_\rho^2 \varepsilon_{\alpha\beta\gamma} \psi^+(x) T_\gamma \psi(x) \right] \delta^{(3)}(\vec{x}-\vec{x}')$$

where in the last line we have taken advantage of the "Jacobi identity"

$$\varepsilon_{\alpha\beta\gamma} \varepsilon_{\gamma\delta\epsilon} + \varepsilon_{\beta\delta\gamma} \varepsilon_{\gamma\alpha\epsilon} + \varepsilon_{\delta\alpha\gamma} \varepsilon_{\gamma\beta\epsilon} = 0 .$$

So

$$m_\rho^4 [\rho_4^\alpha(x), \rho_4^\beta(x')]_{x_0 = x'_0} = - \varepsilon_{\alpha\beta\gamma} f_\rho m_\rho^2 \rho_4^\gamma(x) \delta^{(3)}(\vec{x}-\vec{x}').$$

Our next task is to rewrite the commutation relations among the ρ_μ^α's using C F I. We get

$$[j_k^\alpha(x), j_\ell^\beta(x')]_{x_0 = x'_0} = 0$$

$$[j_0^\alpha(x), j_0^\beta(x')]_{x_0 = x'_0} = i \varepsilon_{\alpha\beta\gamma} j_0^\gamma(x) \delta^{(3)}(\vec{x}-\vec{x}')$$

$$[j_0^\alpha(x), j_\ell^\beta(x')]_{x_0 = x'_0} = i \varepsilon_{\alpha\beta\gamma} j_\ell^\gamma(x) \delta^{(3)}(\vec{x}-\vec{x}') - i \delta_{\alpha\beta} (m_\rho^2 / f_\rho^2) \partial_\ell \delta^{(3)}(\vec{x}-\vec{x}').$$

These commutation relations are seen to be the same as the commutation relations in the quark-field model with the following important exceptions:

i) The space components of j_μ^α commute.

ii) The coefficient of the Schwinger term is an explicitly known, finite c-number.

We have ignored throughout the problem of renormalization. However, from the relation

$$\frac{m_\rho^2}{f_\rho} = \frac{m_\rho^{(0)2}}{f_\rho^{(0)}} \frac{1}{Z_3^{1/2}}$$

derived earlier, we see that the combination $(m_\rho^2/f_\rho)\rho_\mu^\alpha$ is "renormalization invariant". Hence the commutation relations for the currents we have derived still hold even when the ρ field is renormalized.

The finiteness of the Schwinger term can be tested experimentally by looking at the high-energy behavior of colliding beam processes $e^+ + e^- \rightarrow$ hadrons. In Chapter 2 we remarked that because of the Goto-Imamura argument we can express the coefficient of the Schwinger term as an integral over the spectral function. In the gauge field algebra we get

$$\frac{m_\rho^2}{f_\rho^2} = \int \frac{\rho^{(3)}(m^2)}{m^2} dm^2$$

where

$$(\delta_{\mu\nu} - \frac{p_\mu p_\nu}{p^2})\rho^{(3)}(-p^2) = (2\pi)^3 \sum_n \delta^{(4)}(p-p_n) \langle 0|j_\mu^3|n\rangle\langle n|j_\nu^3|0\rangle .$$

Meanwhile, it is straightforward to show that the total cross section for

$e^+ + e^- \rightarrow T=1$ hadronic system

can be written as

$$\sigma^{(T=1)} = (2\pi)^4 \frac{e^4}{4S^2} \sum_n \delta^4(p-p_n) |\langle n|(j_1^3 + i j_2^3)|0\rangle|^2$$

where the 3rd axis is taken in the incident beam direction. We therefore have

$$\sigma^{(T=1)} = \frac{16\pi^3 \alpha^2}{S^2} \rho(m^2),$$

and the gauge field algebra expression for the Schwinger term gives the sum rule

$$\frac{m_\rho^2}{f_\rho^2} = \frac{1}{16\pi^3\alpha^2}\int s\,\sigma^{(T=1)}ds\,.$$

This is a very stringent result. It implies among other things that the hadronic cross section in e^+e^- collisions must go to zero faster than $1/s^2$ (Dooher '67). In contrast, in the quark-field algebra where the Schwinger term is infinite, the above integral is expected to be linearly divergent, which means that $\sigma^{(T=1)}$ is expected to go like $1/s$ (Bjorken '66).

In this section we discussed gauge field algebra corresponding to SU (2) symmetry. It is straightforward to extend our considerations to other higher symmetries such as chiral SU(2) $\otimes$ SU(2).

ω-φ Complex

In addition to a T=1 vector meson coupled to isospin (ρ), the original 1960 paper also proposed the existence of two Y=T=0 vector mesons coupled to hypercharge and baryonic charge. Since they have the same quantum numbers (as far as isospin, hypercharge and J^{PG} are concerned), in the broken eightfold way, the actual mass eigenstates—ω and φ— are expected to be linear combinations of the two pure states coupled to Y and B.

We first define the renormalized φ and ω fields by

$$\langle 0|\varphi_\mu|\omega\rangle = \langle 0|\omega_\mu|\varphi\rangle = 0 .$$

In analogy with

$$f_\rho Q^\alpha = \int d^3x\, J_0^{(\rho),\alpha} ,$$

we expect the Y and B operators to be linear combinations of the space integrals of $J_0^{(\omega)}$ and $J_0^{(\varphi)}$ (sources of ω and φ).

$$f_Y Y = \int d^3x \left[\cos\Theta_Y J_0^{(\varphi)} - \sin\Theta_Y J_0^{(\omega)}\right]$$

$$f_B B = \int d^3x \left[\sin\Theta_B J_0^{(\varphi)} + \cos\Theta_B J_0^{(\omega)}\right],$$

where in general $\Theta_Y \neq \Theta_B \neq 0$, a point emphasized particularly by Kroll, Lee and Zumino ('67). Furthermore, the analogues of

$$j_\mu^\alpha = \frac{m_\rho^2}{f_\rho} \rho_\mu^\alpha \qquad \text{(C F I)}$$

are:

$$2\, j_\mu^{(\text{isoscalor e.m.})} = j_\mu^{(Y)} = \frac{1}{f_Y}\left(\cos\Theta_Y m_\varphi^2 \varphi_\mu - \sin\Theta_Y m_\omega^2 \omega_\mu\right)$$

$$j_\mu^{(B)} = \frac{1}{f_B}\left(\sin\Theta_B m_\varphi^2 \varphi_\mu + \cos\Theta_B m_\omega^2 \omega_\mu\right),$$

where in SU(3) notation

$$j_\mu^{(Y)} = \frac{2}{\sqrt{3}} j_\mu^8 \quad , \quad j_\mu^{(B)} = \sqrt{\frac{2}{3}}\, j_\mu^0 .$$

We note that the electromagnetic current can now be written as

$$j_\mu^{em} = j_\mu^3 + \frac{1}{2} j_\mu^{(Y)}$$

$$= \frac{m_\rho^2}{f_\rho} \rho_\mu^3 + \frac{1}{2f_Y}\left(m_\varphi^2 \cos\Theta_Y \varphi_\mu - m_\omega^2 \sin\Theta_Y \omega_\mu\right).$$

$\cos\Theta_Y/f_Y$ and $\sin\Theta_Y/f_Y$ can be determined from $\omega, \varphi \to e^+ + e^-$ simply by making the following replacements in the formula for $\rho^0 \to e^+e^-$,

$$\Gamma(\rho^0 \to e^+e^-) \approx \frac{(e^2/4\pi)^2}{f_\rho^2/4\pi}\frac{m_\rho}{3}:$$

for $\varphi \to e^+e^-$ substitute $m_\rho \to m_\varphi$, $1/f_\rho \to \cos\Theta_Y/2f_Y$,

for $\omega \to e^+e^-$ substitute $m_\rho \to m_\omega$, $1/f_\rho \to \sin\Theta_Y/2f_Y$.

The angle Θ_Y is directly measureable since (Dashen and Sharp '63)

$$\frac{\Gamma(\omega \to e^+e^-)}{\Gamma(\varphi \to e^+e^-)} = \frac{m_\omega}{m_\varphi}\tan^2\Theta_Y .$$

Θ_B, however, cannot be measured in this way since there are no weak fields coupled to $j_\mu^{(B)}$.

Look now at the "inverse" equations:

$$m_\varphi^2 \varphi_\mu = \frac{1}{\cos(\Theta_Y - \Theta_B)}\left[f_Y \cos\Theta_B\, j_\mu^{(Y)} + f_B \sin\Theta_Y\, j_\mu^{(B)}\right]$$

$$m_\omega^2 \omega_\mu = \frac{1}{\cos(\Theta_Y - \Theta_B)}\left[-f_Y \sin\Theta_B\, j_\mu^{(Y)} + f_B \cos\Theta_Y\, j_\mu^{(B)}\right].$$

The first leads to the matrix element relation,

$$\frac{m_\varphi^2}{m_\varphi^2 - t}\langle B|J_\mu^{(\varphi)}|A\rangle = \frac{1}{\cos(\Theta_Y - \Theta_B)}\left[f_Y\cos\Theta_B\langle B|j_\mu^{(Y)}|A\rangle + f_B\sin\Theta_Y\langle B|j_\mu^{(B)}|A\rangle\right]$$

where $\langle B|j_\mu^{(Y)}|A\rangle$ and $\langle B|j_\mu^{(B)}|A\rangle$ are known exactly at $t=0$. Consider, for example, $\varphi \to K^+K^-, K^0\bar{K}^0$.

At zero momentum transfer, the off-mass-shell $\varphi K\bar{K}$ coupling constant is given by

$$\frac{f_Y\cos\Theta_B}{\cos(\Theta_Y - \Theta_B)}.$$

Assuming that the vector meson form factor (from $\langle B|J_\mu^{(\varphi)}|A\rangle$) doesn't vary much between $t=0$ and $t=m_\varphi^2$, we obtain from the observed φ width

$$\Gamma(\varphi \to K^+K^-) = 1.7 \pm 0.4\,\text{Mev},$$

$$\frac{1}{4\pi}\left[\frac{f_Y\cos\Theta_B}{\cos(\Theta_Y - \Theta_B)}\right]^2 = 1.4 \pm 0.3 .$$

As another example consider the $\varphi N\bar{N}$ coupling, which seems to be very small. (No backward peak in $K^- + p \rightarrow \Lambda + \varphi$ interpretable as N exchange.) We get (again assuming slow variation of the vector meson form factor)

$$f_Y \cos\theta_B \approx - f_B \sin\theta_Y .$$

At zero momentum transfer the diagrams

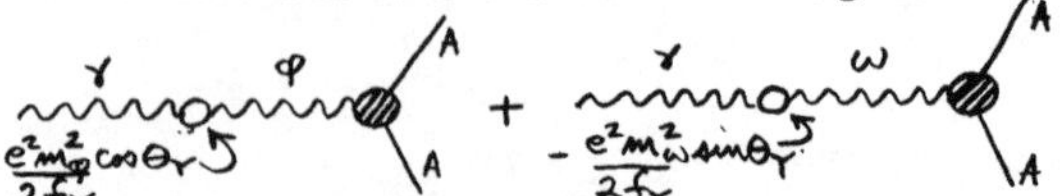

$$\frac{e^2 m_\varphi^2}{2f_Y}\cos\theta_Y \qquad - \frac{e^2 m_\omega^2 \sin\theta_Y}{2f_Y}$$

should give the isoscalar part of the electric charge of A (= $Y^{(A)}/2$). It is easy to show that our formalism satisfies this requirement automatically.

So far everything we have said is completely independent of SU(3). Exact SU(3) implies:

(i) $\theta_Y = \theta_B = 0 \left\{ \begin{array}{l} \varphi \text{ is pure octet} \\ \omega \text{ is pure singlet} \end{array} \right.$ (or vice-versa),

(ii) $f_Y = \frac{\sqrt{3}}{2} f_\rho$ (from the C.G.coeff.for pure F-type coupling).

In broken SU(3), $\theta_B \neq 0$, $\theta_Y \neq 0$, and in general $\theta_B \neq \theta_Y$ — the actual values depending on how SU(3) is broken.

Briefly consider two special models of ω-φ mixing:

(i) <u>The Mass Mixing Model</u> (Okubo '63;Sakurai '63)

$$\mathcal{L}_{sym} = -\frac{1}{4}\mathrm{Tr}(V_{\mu\nu}V_{\mu\nu}) - \frac{1}{2}m_8^2\,\mathrm{Tr}(V_\mu V_\mu) - \frac{1}{4}\omega^{(1)}_{\mu\nu}\omega^{(1)}_{\mu\nu} - \frac{1}{2}m_1^2\,\omega^{(1)}_\mu \omega^{(1)}_\mu ,$$

$$\mathcal{L}_{\text{breaking}} = -\,\delta m^2\,\mathrm{Tr}(V_\mu \lambda_8 V_\mu) - m_{\omega\varphi}^2\,\omega_\mu^{(1)}\omega_\mu^{(8)}.$$

In this model the 1st order breaking is in the mass (δm^2) term. The last term is the ω-φ mixing term, in the absence of which the vector octet masses would satisfy the Gell-Mann-Okubo relation,

$$m_{K^*}^2 = \frac{1}{4}\left(m_\rho^2 + 3m_{\omega^{(8)}}^2\right).$$

Diagonalize $\mathcal{L}$ to eliminate the mixing term, we get

$$\frac{1}{3}(4m_{K^*}^2 - m_\rho^2) = m_\varphi^2 \cos^2\theta + m_\omega^2 \sin^2\theta,$$

$$\theta = \Theta_B = \Theta_Y = 39°.$$

(ii) <u>The Current Mixing Model</u> (Coleman and Schnitzer '64)

Here the 1st order breaking is in the "kinetic" ($V_{\mu\nu}V_{\mu\nu}$) term.

$$\mathcal{L}_{\text{breaking}} = -\,\lambda\,\mathrm{Tr}(V_{\mu\nu}\lambda_8 V_{\mu\nu}) - \lambda_{\omega\varphi}\,\omega_{\mu\nu}^{(1)}\omega_{\mu\nu}^{(8)}$$

In the absence of ω-φ mixing, the vector octet masses would satisfy

$$\frac{1}{m_{K^*}^2} = \frac{1}{4}\left(\frac{1}{m_\rho^2} + \frac{3}{m_{\omega^{(8)}}^2}\right).$$

To show this first note that for each component of the vector octet we have

$$-\frac{1}{4}(1+\varepsilon)\,v_{\mu\nu}v_{\mu\nu} - \frac{1}{2}m_0^2\,v_\mu v_\mu,$$

where

$$\varepsilon_\rho : \varepsilon_{K^*} : \varepsilon_{\omega^{(8)}} = -2 : 1 : 2.$$

Redefining

$$v_\mu' = \sqrt{1+\varepsilon}\;v_\mu,$$

we get the renormalized mass

$$m^{*2} = m_0^2/(1+\varepsilon).$$

Since it is ε that satisfies a Gell-Mann-Okubo like relation, we must use $1/m^2$ instead of m^2 in the mass formula. Diagonalizing as before and appropriately renormalizing the vector fields, one can show

$$\frac{1}{3}\left(\frac{4}{m_{K^*}^2} - \frac{1}{m_\rho^2}\right) = \left(\frac{\cos^2\theta}{m_\varphi^2} + \frac{\sin^2\theta}{m_\omega^2}\right),$$

$$m_\omega m_\varphi \tan\theta = m_\omega^2 \tan\theta_Y = m_\varphi^2 \tan\theta_B ,$$

$$\theta = 26^\circ , \ \theta_Y = 33^\circ , \ \theta_B = 21^\circ .$$

In Chapter 6 we'll discuss how spectral function sum rules may be used to relate f_ρ and f_Y in the <u>broken</u> eightfold way and present an argument in favor of the current mixing model.

CHAPTER IV. PCAC AND THE GOLDBERGER-TREIMAN RELATION

Matrix Elements for $\pi^{\pm}$ decay and Neutron Decay

In this chapter we explore the connection between the matrix elements of $\pi^{\pm}$ decay and neutron β decay (the axial-vector part). The most general form of the matrix element for the weak decay of the pion is

$$\langle 0 | j^{i\alpha}_{5\mu}(0) | \pi^{\alpha} \rangle = \frac{i\, C_\pi}{\sqrt{2\omega}}\, p_\mu \,.$$

This follows just from Lorentz invariance and parity.

For $\pi^- \rightarrow e^- + \bar{\nu}$ we have

$$\langle 0 | j^{1+i2}_{5\mu}(0) | \pi^- \rangle = \frac{i\sqrt{2}}{\sqrt{2\omega}}\, C_\pi\, p_\mu \,,$$

which corresponds to the effective Lagrangian

$$i \frac{G}{\sqrt{2}} \cos\theta\, \sqrt{2}\, C_\pi\, \partial_\mu \pi^-\, \bar{e}\gamma_\mu (1+\gamma_5)\nu \,.$$

Experimentally

$$C_\pi = 94 \text{ MeV}.$$

(Other definitions of the π decay constants are: Weinberg's $F_\pi = 2\,C_\pi$ and Gell-Mann's $f_\pi = 1/2\, C_\pi$.)

For the axial-vector part of neutron decay we have

$$\langle N' | j^{i\alpha}_{5\mu}(0) | N \rangle = \sqrt{\frac{m_N^2}{E'E}} \left[i\,\bar{u}'\gamma_\mu\gamma_5 \frac{\tau_\alpha}{2} u\, F_A(t) + q_\mu\, \bar{u}'\gamma_5 \frac{\tau_\alpha}{2} u\, F_P(t) \right],$$

$$q_\mu = (p' - p)_\mu$$

where the $\bar{u}'\gamma_5\tau_\alpha u$ term is known as the induced pseudoscalar term. Note that a term of the form $q_\nu \bar{u}'\sigma_{\mu\nu}\gamma_5\tau_\alpha u$, though allowed by parity, is forbidden by G parity. (This is an example of what Weinberg calls "second class currents".)

In the Lagrangian language such a term would correspond to $\partial_\mu(\bar{\psi}\sigma_{\mu\nu}\gamma_5\tau_\alpha\psi)$ which has G parity opposite to $i\bar{\psi}\gamma_\mu\gamma_5\tau_\alpha\psi$

and $i\partial_\mu(\bar{\psi}\gamma_5\tau_\alpha\psi)$. It is also ruled out by time reversal (see e.g. J.S. Bell's Les Houches lectures), but this is equivalent to G parity by CPT and isospin. Experimentally

$$F_A(0) = -g_A/g_V = 1.18 \pm 0.02.$$

<u>Troubles with CAC</u> (conserved axial-vector current).

In the vector part of the semileptonic weak interactions the conserved vector current hypothesis of Feynman and Gell-Mann led to many successful results. So we may be tempted to postulate that the axial-vector current that appears in the weak interactions is also divergenceless. Exact axial-vector conservation, however, leads to a contradiction with experiment.

Suppose

$$\partial_\mu j^{\alpha}_{5\mu} = 0.$$

Then taking the divergence of the pion decay matrix element, we get

$$0 = \langle 0|\partial_\mu j^\alpha_{5\mu}(x)|\pi^\alpha\rangle = ip_\mu\Big[ic_\pi p_\mu \frac{e^{ip\cdot x}}{\sqrt{2\omega}}\Big] = c_\pi m_\pi^2 \frac{e^{ip\cdot x}}{\sqrt{2\omega}}.$$

So the CAC hypothesis requires (J.C.Taylor, '58)

(i) $m_\pi^2 = 0$ and/or

(ii) $c_\pi = 0$.

But both of these are clearly false; $m_\pi \neq 0$, and $\pi^\pm$ decay does occur!

CAC also leads to trouble with neuton decay.

$$\langle N'|\partial_\mu j^\alpha_{5\mu}|N\rangle = 0$$

implies

$$F_A(t)(-2m_N)\bar{u}'\gamma_5\frac{\tau_\alpha}{2}u + q^2 F_P(t)\bar{u}'\gamma_5\frac{\tau_\alpha}{2}u = 0,$$

where we have used

$$(i\gamma\cdot p + m_N)u = 0 \quad , \quad \bar{u}'(i\gamma\cdot p' - m_N) = 0 \quad , \quad q = p' - p .$$

Hence

$$F_P(t) = \frac{2m_N}{q^2} F_A(t)$$

$$\langle N'|j^\alpha_{5\mu}(0)|N\rangle = \sqrt{\frac{m_N^2}{EE'}}\, F_A(t)\left[i\bar{u}'\gamma_\mu\gamma_5\frac{\tau_\alpha}{2}u + 2m_N\frac{q_\mu}{q^2}\bar{u}'\gamma_5\frac{\tau_\alpha}{2}u\right].$$

In the non-relativistic limit $(|\vec{p}|, |\vec{p}'| << m_N)$

$$i\bar{u}'\gamma_k\gamma_5 u \overset{N.R.}{=} -\chi'\sigma_k\chi ,$$

$$\bar{u}'\gamma_5 u \overset{N.R.}{=} \chi'\frac{\vec{\sigma}\cdot\vec{q}}{2m_N}\chi ,$$

so that

$$\langle N'|\vec{j}_5^{\,\alpha}|N\rangle \overset{N.R.}{=} F_A(0)\,\chi'\left[-\vec{\sigma} + \frac{\vec{q}}{|\vec{q}|^2}\vec{\sigma}\cdot\vec{q}\right]\frac{\tau^\alpha}{2}\chi .$$

Thus we would have a huge induced pseudoscalar contribution which would completely cancel (for $\langle\vec{\sigma}\rangle \parallel \vec{q}$) the usual Gamow-Teller term (argument due to Goldberger and Treiman '58). What is worse, $F_P(t) \sim 1/q^2$ would imply a long range interaction between the hadronic and the leptonic currents mediated by a zero mass pseudoscalar meson.

In short, the CAC hypothesis is crazy. All these troubles were already known in 1958, so people stopped thinking about CAC for a while.

Nambu's Derivation of the G-T Relation

Let us return now to the silly formula we got from CAC.

$$\langle N'|\, j^{\alpha}_{5\mu}(0)\,|N\rangle = \sqrt{\frac{m_N^2}{E'E}}\; F_A(t)\left[i\bar{u}'\gamma_\mu\gamma_5\frac{\tau_\alpha}{2}u + 2m_N\frac{q_\mu}{q^2}\,\bar{u}'\gamma_5\frac{\tau_\alpha}{2}u\right]$$

Even though there is no zero mass pseudoscalar meson, there is π — the lightest of the hadrons. So Nambu ('60) postulated that the $1/q^2$ in the second term should be interpreted as

$$\frac{1}{q^2} = \lim_{m_\pi \to 0} \frac{1}{q^2 + m_\pi^2}\,.$$

After all, the idealized world in which $m_\pi^2 = 0$ and $j^{\alpha}_{5\mu}$ is conserved might not be too different from the real world with $m_\pi^2 \neq 0$. In other words, we assume no drastic change as $m_\pi^2 \to 0$. Meanwhile, it was known for some time that $F_p(t)$ has a pion pole contribution due to

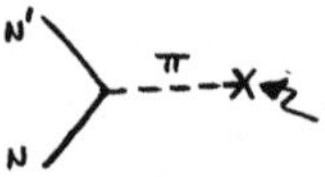

interaction with axial current.

This diagram gives the covariant matrix element,

$$G_{\pi NN}\,\bar{u}'\gamma_5\tau_\alpha u\,\frac{C_\pi q_\mu}{q^2 + m_\pi^2}$$

($G_{\pi NN}$ is defined, without a factor of 1/2, i.e. by $\mathcal{L}_{INT} = i\,G_{\pi NN}\bar{N}\gamma_5\tau_\alpha N\pi^\alpha$), which, according to Nambu, is to be compared with

$$F_A(t)\,2m_N\frac{q_\mu}{q^2 + m_\pi^2}\,\bar{u}'\gamma_5\frac{\tau_\alpha}{2}u.$$

Assuming that $F_A(t)$ is slowly varying between $0 \leq t \leq m_\pi^2$,

$$F_A(0)\,m_N = C_\pi G_{\pi NN}$$

$$C_\pi = \left(-g_A/g_V\right)\frac{m_N}{G_{\pi NN}}\,.$$

This, the Goldberger-Treiman relation, expresses the pion decay constant in terms of g_A/g_V and $G_{\pi NN}$. Note that the pion decay rate is <u>inversely</u> proportional to $G^2_{\pi NN}$ — just the opposite of what one would expect from

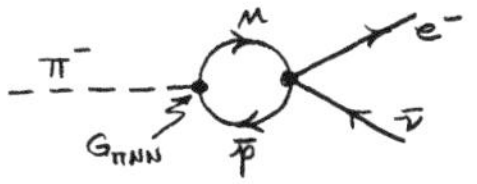

(Goldberger and Treiman ('58) first derived this relation using an entirely different argument based on a dispersion theoretic treatment of the $\pi \to N\bar{N}$ amplitude.). With $G^2_{\pi NN}/4\pi = 14.6$, we obtain

$$c_\pi = 82 \text{ MeV},$$

$$\frac{(c_\pi) \text{ from G-T rel.}}{(c_\pi) \text{ experiment}} = \frac{82}{94} = .87.$$

- good to about 13% in the amplitude.

Let us again take $\langle N'|j^\alpha_{5\mu}|N\rangle$ obtained from CAC and calculate $\langle N'|\partial_\mu j^\alpha_{5\mu}|N\rangle$ this time with $1/(q^2+m_\pi^2)$ in place of $1/q^2$.

$$\langle N'|\partial_\mu j^\alpha_{5\mu}(0)|N\rangle = -i\sqrt{\frac{m_N^2}{E'E}}\, F_A(t)\left[-2m_N \bar{u}'\gamma_5 \frac{\tau_\alpha}{2} u + \frac{q^2\, 2m_N}{(q^2+m_\pi^2)} \bar{u}'\gamma_5 \frac{\tau_\alpha}{2} u\right]$$

$$= \sqrt{\frac{m_N^2}{E'E}}\, F_A(t)\left(\frac{m_N m_\pi^2}{q^2+m_\pi^2}\right) i\,\bar{u}'\gamma_5 \tau_\alpha u$$

But $(\Box - m_\pi^2)\pi^\alpha = -J^{(\pi),\alpha}$ implies

$$\langle N'|\pi^\alpha|N\rangle = \frac{\langle N'|J^{(\pi),\alpha}|N\rangle}{q^2+m_\pi^2} \underset{q^2\approx -m_\pi^2}{\approx} i\sqrt{\frac{m_N^2}{E'E}}\,\frac{G_{\pi NN}\,\bar{u}'\gamma_5\tau_\alpha u}{(q^2+m_\pi^2)}.$$

Assuming that the matrix elements $\langle N'|\partial_\mu j^\alpha_{5\mu}|N\rangle$ and $\langle N'|\pi^\alpha|N\rangle$ vary little between $t=0$ and $t=m_\pi^2$, we get

$$\langle N'|\partial_\mu j^\alpha_{5\mu}(0)|N\rangle \underset{0<t<m_\pi^2}{\approx} (-g_A/g_V)\frac{m_N}{G_{\pi NN}} m_\pi^2 \langle N'|\pi^\alpha|N\rangle$$
$$= C_\pi m_\pi^2 \langle N'|\pi^\alpha|N\rangle,$$

where the G-T relation has been used. This suggests an operator equation (or "divergence - field identity")

$$\partial_\mu j^\alpha_{5\mu} = C_\pi m_\pi^2 \pi^\alpha$$

known as the PCAC relation first written down by Gell-Mann and Lévy ('60). Note that $j^\alpha_{5\mu}$ becomes divergenceless as $m_\pi^2 \rightarrow 0$, as expected.

Models Satisfying PCAC

It is natural to ask whether there are simple Lagrangian models that satisfy the PCAC relation.

(1) Gradient Coupling Model [The connection with the G-T rel., originally due to Feynman, first appeared in Gell-Mann and Lévy('60).]

$$\mathcal{L} = -(\bar{N}\gamma_\mu\partial_\mu N + m_N\bar{N}N) - \frac{1}{2}[(\partial_\mu\pi^\alpha)^2 + (m_\pi\pi^\alpha)^2]$$
$$- \frac{i\,G_{\pi NN}}{2m_N}\partial_\mu\pi^\alpha\,\bar{N}\gamma_\mu\gamma_5\tau_\alpha N,$$

where the derivative coupling in the last term is equivalent (to first order only) to the pseudoscalar coupling, $i\,G_{\pi NN}\pi^\alpha\bar{N}\gamma_5\tau_\alpha N$. If the pion mass were zero, $\mathcal{L}$ would be invariant under

$$\pi^\alpha \rightarrow \pi^\alpha + \varepsilon^\alpha \qquad (\varepsilon^\alpha = \text{constant isovector}).$$

Treating ε^α as a space-time dependent function, the current generated by this "gauge" transformation is

$$\frac{\delta \mathcal{L}}{\delta(\partial_\mu \Xi^\alpha)} = -\partial_\mu \pi^\alpha - i\frac{G_{\pi NN}}{2m_N}\bar{N}\gamma_\mu\gamma_5\tau_\alpha N$$

$$\frac{\delta \mathcal{L}}{\delta \Xi^\alpha} = -m_\pi^2 \pi^\alpha.$$

So from the Gell-Mann-Lévy equation, we get

$$\partial_\mu\left[-\partial_\mu \pi^\alpha - i\frac{G_{\pi NN}}{2m_N}\bar{N}\gamma_\mu\gamma_5\tau_\alpha N\right] = -m_\pi^2 \pi^\alpha,$$

which is also obtainable directly from the equation of motion for the pion field. Thus we have succeeded in constructing an axial current that gets conserved in the $m_\pi^2 \to 0$ limit, and it is natural to identify it with the $j_{5\mu}^\alpha$ that appears in weak interactions — apart from a proportionality factor which can be seen to be $-C_\pi$ if the pion decay is to be given correctly.

$$j_{5\mu}^\alpha = C_\pi \partial_\mu \pi^\alpha + i C_\pi \frac{G_{\pi NN}}{m_N}\bar{N}\gamma_\mu\gamma_5\frac{\tau_\alpha}{2}N,$$

$$\partial_\mu j_{5\mu}^\alpha = C_\pi m_\pi^2 \pi^\alpha$$

Compare this with the matrix element appearing in low momentum transfer nucleon β-decay $(p'=p)$,

$$\langle N'|j_{5\mu}^\alpha(0)|N\rangle = \frac{m_N}{E}F_A(0)\, i\,\bar{u}'\gamma_\mu\gamma_5\frac{\tau^\alpha}{2}u,$$

and we again get the G-T relation.

$$F_A(0) = C_\pi G_{\pi NN}/m_N$$

(2) The σ Model [Gell-Mann and Lévy ('60); partly based on the earlier work of Schwinger ('57)]

In order to construct a strong Lagrangian with pseudoscalar-type pion coupling, we introduce a σ meson having the same quantum numbers as the vacuum $(T=Y=0,$ $J^{PCG}=0^{+++})$.

$$\mathcal{L} = -\bar{N}\gamma_\mu\partial_\mu N - \frac{1}{2}\left[(\partial_\mu\pi^\alpha)^2 + (\partial_\mu\sigma)^2 + m_\pi^2(\pi^\alpha\pi^\alpha + \sigma^2)\right]$$
$$+ G_{\pi NN}(i\pi^\alpha\bar{N}\gamma_5\tau^\alpha N + \sigma\bar{N}N)$$
$$- \lambda\left(\pi^\alpha\pi^\alpha + \sigma^2 - \frac{m_N^2}{G_{\pi NN}^2}\right)^2 - \frac{m_N m_\pi^2}{G_{\pi NN}}\sigma$$

What, no nucleon mass term and $m_\sigma = m_\pi$? But consider the "translation"

$$\sigma \to \sigma - \frac{m_N}{G_{\pi NN}},$$

then the free nucleon Lagrangian takes the usual form $-(\bar{N}\gamma_\mu\partial_\mu N + m_N\bar{N}N)$. Moreover, the ($\sigma$ mass)^2 becomes

$$m_\sigma^2 = m_\pi^2 + \frac{8m_N^2\lambda}{G_{\pi NN}^2}$$

so the σ and π need not be degenerate. Coming back to the original form of the Lagrangian, let us perform the "gauge transformation",

$$N \to (1 + i\,\varepsilon_\alpha\gamma_5\frac{\tau_\alpha}{2})N$$
$$\pi^\alpha \to \pi^\alpha - \varepsilon^\alpha\sigma$$
$$\sigma \to \sigma + \varepsilon^\alpha\pi^\alpha.$$

(The $T=1$, $J^P = 0^-$ π and the $T=0$, $J^P = 0^+$ σ form a "chiral quadruplet".)

We obtain

$$\frac{\delta\mathcal{L}}{\delta(\partial_\mu\varepsilon_\alpha)} = -i\bar{N}\gamma_\mu\gamma_5\frac{\tau_\alpha}{2}N + \sigma\partial_\mu\pi^\alpha - \pi^\alpha\partial_\mu\sigma$$

$$\frac{\delta\mathcal{L}}{\delta\varepsilon_\alpha} = -\frac{m_N m_\pi^2}{G_{\pi NN}}\pi^\alpha.$$

(Only the last term in $\mathcal{L}$ breaks the symmetry under the gauge transformation.) Thus we again have an axial current that gets conserved in the $m_\pi^2 \to 0$ limit.

$$j_{5\mu}^\alpha \equiv \frac{g_A}{g_V}\frac{\delta \mathcal{L}}{\delta(\partial_\mu \varepsilon_\alpha)} = -\frac{g_A}{g_V} i \bar{N}\gamma_\mu \gamma_5 \frac{\tau_\alpha}{2} N + \frac{g_A}{g_V}(\sigma \partial_\mu \pi^\alpha - \pi^\alpha \partial_\mu \sigma)$$

$$\partial_\mu j_{5\mu}^\alpha = -\frac{g_A}{g_V}\frac{m_N}{G_{\pi NN}} m_\pi^2 \pi^\alpha$$

There are other Lagrangian models which give PCAC (e.g. Gürsey's nonlinear model), but we won't discuss them in this course.

PCAC and Pion-Pole Dominance (Bernstein, Fubini, Gell-Mann and Thirring '60)

When taken between the vacuum and a single π state, PCAC directly gives the π decay constant - by definition.

$$\langle 0 | \partial_\mu j_{5\mu}^\alpha(x) | \pi^\alpha \rangle = C_\pi m_\pi^2 \frac{e^{ip\cdot x}}{\sqrt{2\omega}}$$

Other matrix elements are not as trivial.

$$\langle B | \partial_\mu j_{5\mu}^\alpha | A \rangle = \frac{C_\pi m_\pi^2}{m_\pi^2 - t} \langle B | J^{(\pi),\alpha} | A \rangle$$

(This looks very much like pion-pole dominance of $\partial_\mu j_{5\mu}^\alpha$.)

Between nucleon states at $t = 0$,

$$C_\pi \langle N' | J^{(\pi),\alpha}(0) | N \rangle_{t=0} = \langle N' | \partial_\mu j_{5\mu}(0) | N \rangle_{t=0}$$

$$= i\sqrt{\frac{m_N^2}{E'E}}\, F_A(0)\, 2m_N \bar{u}' \gamma_5 \frac{\tau_\alpha}{2} u .$$

If this were evaluated at $t = m_\pi^2$, then we could use the πNN coupling constant on the LHS.

$$\langle N' | J^{(\pi),\alpha}(0) | N \rangle_{t=m_\pi^2} = i\, G_{\pi NN} \sqrt{\frac{m_N^2}{E'E}}\, \bar{u}' \gamma_5 \tau_\alpha u$$

As we go from $t = m_\pi^2$ to $t = 0$, a correction factor is needed - namely the πNN vertex form factor, $K_{\pi NN}(t)$, which is normalized so that $K_{\pi NN}(m_\pi^2) = 1$. Departure from the exact G-T rel. is due to this factor.

$$m_N F_A(0) = C_\pi G_{\pi NN} K_{\pi NN}(0) \quad \text{(exact!)}$$

As for complete pion-pole dominance of the divergence of the axial current, i.e.

$$\langle B | \partial_\mu j^\alpha_{5\mu} | A \rangle \propto 1/(m_\pi^2 - t),$$

we must have $\langle B | J^{(\pi),\alpha} | A \rangle = \text{constant} = \langle B | J^{(\pi),\alpha} | A \rangle_{t = m_\pi^2}$.

In other words, the πAB form factor must be constant.

To make a more quantitative comparison with the dispersion theory method, start with the most general matrix element for $\langle N' | j^\alpha_{5\mu} | N \rangle$ and take the divergence.

$$\langle N' | \partial_\mu j^\alpha_{5\mu} | N \rangle = (-i q_\mu) \sqrt{\frac{m_N^2}{E'E}} \left[i \bar{u}' \gamma_\mu \gamma_5 \frac{\tau_\alpha}{2} u F_A(t) + q_\mu \bar{u}' \gamma_5 \frac{\tau_\alpha}{2} u F_P(t) \right]$$

$$= \sqrt{\frac{m_N^2}{E'E}} \left(2 m_N F_A(t) + t F_P(t) \right) i \bar{u}' \gamma_5 \frac{\tau_\alpha}{2} u$$

Write a dispersion relation for the combination $2 m_N F_A(t) + t F_P(t)$, where only $T = 1$, $J^{PG} = 0^{--}$ states contribute (π, 3π, 5π, $K\bar{K}$, $N\bar{N}$, etc.).

Assuming the dispersion integral needs no substractions,

$$2 m_N F_A(t) + t F_P(t) = \frac{R_{\pi\,pole}}{m_\pi^2 - t} + \int_{(3m_\pi)^2}^{\infty} \frac{t' \sigma(t') dt'}{t' - t}$$

where $R_{\pi\,pole}$ is the residue of the pion pole. To calculate $R_{\pi\,pole}$, start with the Feynman graph contribution for the pion exchange

$$-\sqrt{\frac{m_N^2}{E'E}} (-i q_\mu)(G_{\pi NN} i \bar{u}' \gamma_5 \tau_\alpha u) \left(\frac{i C_\pi q_\mu}{q^2 + m_\pi^2 - i\varepsilon} \right),$$

evaluate it at the pion pole ($q^2 = -m_\pi^2$), and compare with the general formula for $\langle N' | \partial_\mu j^{\alpha}_{5\mu} | N \rangle$.

$$R_{\pi\,\text{pole}} = 2\, C_\pi\, G_{\pi NN}\, m_\pi^2$$

$$2\, m_N F_A(0) + t\, F_p(t) = \frac{2\, C_\pi\, G_{\pi NN}\, m_\pi^2}{m_\pi^2 - t} + \text{continuum}$$

(So far no approximations have been made.) Evaluate at $t = 0$.

$$2 m_N F_A(0) = 2 C_\pi G_{\pi NN} + \int_{(3m_\pi)^2}^{\infty} \sigma(t')\,dt'$$

If we ignore the continuum, then we again get the G-T relation. No one doubts that near the π pole, only the pion pole contribution is important. But the validity of the G-T relation depends on π pole dominance of the divergence of the axial current even at $t = 0$.

By this time you may have noticed the following analogy between ρ and π couplings:

$j^{\alpha}_{\mu} = \frac{m_\rho^2}{f_\rho} \rho^{\alpha}_{\mu}$ (current-field identity)	$\partial_\mu j^{\alpha}_{5\mu} = C_\pi m_\pi^2 \pi^{\alpha}$ (divergence-field identity)
f_ρ^{-1} measurable from $\rho \to e^+e^-$	C_π measurable from $\pi^+ \to \mu^+\nu'$
$f_{\rho\pi\pi} = f_\rho$	G-T relation
Validity of " depends on:	Validity of " depends on:
i) slow variation of $F_{\rho\pi\pi}(t)$ between $t = 0$ and m_ρ^2 (CFI language)	i) slow variation of $K_{\pi NN}(t)$ between $t = 0$ and m_π^2 (DFI language)
ii) ρ dominance of π e.m. form factor (dispersion theory language)	ii) π dominance of divergence of axial current (dispersion theory language)

iii) ρ coupled to isospin at $t=0$ with strength f_ρ.

iii) π coupled to the divergence of the axial current at $t=0$ with strength $1/c_\pi$.

PCAC and SU(3); Generalized G-T Relations

If we apply the Nambu argument to the axial matrix elements appearing in the β decay of strange baryons, we get

$$\langle B|\, j_{5\mu}^{1\pm i2,\,4\pm i5}\,|A\rangle = \sqrt{\frac{m_A m_B}{E_A E_B}}\, F_A^{(BA)}(t)\, \bar{u}_B\Big(i\gamma_\mu\gamma_5 + \frac{m_B+m_A}{q^2+m_{\pi,K}^2}\gamma_5 q_\mu\Big)u_A .$$

Note that for $|\Delta S|=1$ processes, a K pole appears in place of the π pole. The SU(3) version of the G-T relation is

$$F_A^{(BA)}(0)\,(m_B+m_A) = \sqrt{2}\left\{\begin{array}{l} C_\pi G_{\pi^+ AB} \\ C_K G_{K^+AB} \end{array}\right. ,$$

where $\sqrt{2}$ appears because we have used $1\pm i2,\ 4\pm i5$, not α. Since $G_{\pi^+ pn} = \sqrt{2}\, G_{\pi NN}$, the familiar form of the G-T relation is obtained as a special case. In the Cabibbo theory, the $F_A^{(BA)}(0)$ have both F and D-type contributions, and are related by SU(3) C-G coefficients (Cabibbo '63).

	$\bar{B}$ A	$F_A^{(BA)}(0)$
$\Delta S=0$	$\bar{p}n$	D + F = 1.18
	$\bar{\Lambda}\Sigma^\pm$	$\sqrt{\frac{2}{3}}$ D
$\Delta S=-1$	$\bar{p}\Lambda$	$-(D+3F)/\sqrt{6}$
	$\bar{n}\Sigma^-$	D-F
	$\bar{\Lambda}\Xi^-$	$-(D-3F)/\sqrt{6}$

To the extent that the baryon masses are degenerate and $c_\pi = c_K$, the generalized G-T relation implies

$$(D/F)_{\text{axial } \beta \text{ decay}} = (D/F)_{\text{p.s. meson coupling}} .$$

The cleanest way of determining the D/F ratio is to use $\Sigma^{\pm} \rightarrow \Lambda + e^{\pm} + \nu$ (essentially pure A. By CVC the vector part gives no contribution at $t = 0$.). Recently the Maryland group found

$$\alpha = \frac{D}{D+F} = 0.67 \pm 0.03,$$

which is consistent with the α for pseudoscalar meson coupling determined from KN dispersion relations.

For $|\Delta S| = 1$ processes the G-T relation - or the concept of K pole dominance - may not be so good. The pole at m_K^2 is not far removed from the continuum starting at $(m_K + m_\pi)^2$. Nevertheless, we may still write

$$\partial_\mu j_{5\mu}^{\alpha} = C_K m_K^2 K^\alpha \qquad \alpha = 4,5,6,7.$$

In the Cabibbo theory, the actual decay rate for K has a $\sin^2\theta_A$ factor.

$$\frac{\Gamma(K^+ \rightarrow \mu^+ \nu')}{\Gamma(\pi^+ \rightarrow \mu^+ \nu')} = \left(\frac{C_K}{C_\pi}\right)^2 \tan^2\theta_A \left(\frac{m_K}{m_\pi}\right) \left[\frac{1 - m_\mu^2/m_K^2}{1 - m_\mu^2/m_\pi^2}\right]^2$$

If we assume (strict SU(3) limit) $C_K = C_\pi$, then

$$\sin\theta_A = 0.264 \pm 0.001.$$

Comparison of the vector parts of neutron β decay and μ decay measures $\cos^2\theta_V = 1 - \sin^2\theta_V$. From this

$$\sin\theta_V = 0.208 \pm 0.007.$$

Explanations for this discrepancy are:

i) $\theta_A \neq \theta_V$ (This is ugly!)

ii) $\theta_A = \theta_V$ but

$\frac{C_K}{C_\pi} \approx 1.3$ due to SU(3) breaking.

CHAPTER V. SOFT PION PROCESSES

Review of the Reduction Technique

We need formulas for S matrix elements for pion emission, absorption, and scattering. So assuming that you are somewhat familiar with quantum field theory in the Heisenberg representation, we quickly derive (non-rigorously) the so-called reduction formulas (derived by Low, Lehmann, Symanzik and Zimmermann and others, '55.)

Consider the neutral scalar field, $\varphi(x)$, in the Heisenberg representation (it is straightforward to generalize to the more realistic pion field with isospin.) We know very little about $\varphi(x)$, just:

$(\Box - m^2)\varphi(x) = -J(x)$ (nature of source unspecified),

$[\varphi(x), \varphi(x')] = 0$ for space-like separation,

$\lim_{t \to \mp\infty} \langle B|\varphi|A\rangle = \langle B|\varphi^{in,out}|A\rangle$ (asymptotic condition).

(This discussion ignores the renormalization problem, i.e. the $\sqrt{Z}$ factor. See Bjorken and Drell II for details.). $\varphi^{in,out}$ satisfy the free field equation,

$$(\Box - m^2)\varphi^{in,out} = 0,$$

and the free field commutation relation,

$$[\varphi^{in,out}_{(x)}, \varphi^{in,out}_{(x')}] = i\Delta(x-x').$$

So we may consider the usual free field expansion for $\varphi^{in,out}$

$$\varphi^{in,out}(x) = \sum_{\mathbf{k}}[a_{\mathbf{k}}^{in,out} f_{\mathbf{k}}(x) + (a_{\mathbf{k}}^{in,out})^{+} f_{\mathbf{k}}^{*}(x)].$$

$$f_{\vec{k}} = \frac{e^{ik\cdot x}}{\sqrt{2\omega V}}, \quad f^*_{\vec{k}} = \frac{e^{-ik\cdot x}}{\sqrt{2\omega V}}, \quad \begin{array}{l} k = (\vec{k}, i\omega) \\ \omega = \sqrt{\vec{k}^2 + m^2} \end{array}$$

(Actually, it is better to use wave-packets here.) The f's satisfy the orthogonality conditions,

$$i\int d^3x \begin{Bmatrix} f_{\vec{k}'} \overleftrightarrow{\frac{\partial}{\partial x_0}} f_{\vec{k}} \\ f^*_{\vec{k}'} \overleftrightarrow{\frac{\partial}{\partial x_0}} f_{\vec{k}} \\ f^*_{\vec{k}'} \overleftrightarrow{\frac{\partial}{\partial x_0}} f^*_{\vec{k}} \end{Bmatrix} = \begin{Bmatrix} 0 \\ \delta_{\vec{k}'\vec{k}} \\ 0 \end{Bmatrix}$$

where

$$A(x) \overleftrightarrow{\frac{\partial}{\partial x_0}} B(x) \equiv A \frac{\partial B}{\partial x_0} - \frac{\partial A}{\partial x_0} B .$$

From these it follows that

$$a_{\vec{k}}^{\text{in, out}} = i\int d^3x \, f^*_{\vec{k}} \overleftrightarrow{\frac{\partial}{\partial x_0}} \varphi^{\text{IN, OUT}}$$

$$(a_{\vec{k}}^{\text{in, out}})^\dagger = i\int d^3x \, \varphi^{\text{IN, OUT}} \overleftrightarrow{\frac{\partial}{\partial x_0}} f_{\vec{k}} .$$

For a single particle state we have

$$\langle 0|\varphi(x)|k\rangle = \langle 0|\varphi^{in}(x)|k\rangle .$$

This is derivable from

$$\varphi(x) = \varphi^{in}(x) + \int d^4x' \Delta_{RET}(x - x') J(x')$$

and

$$\langle 0|J(x)|k\rangle = -(\Box - m^2)\, e^{ik\cdot x} \langle 0|\varphi(0)|k\rangle = 0 .$$

So even though $\varphi(x)$ does <u>not</u> satisfy the free K-G equation, we still have

$$\langle 0|\varphi(x)|k\rangle = \frac{e^{ik\cdot x}}{\sqrt{2\omega V}},$$

which we have already assumed earlier in the course.

Of particular interest is the matrix element for the

absorption of a single pion

$$A + \pi_{\mathbf{k}} \to B ,$$

which we now derive. First

$$\langle B|A\pi_{\mathbf{k}}\rangle = \langle B|a_{\mathbf{k}}^{in\,+}|A\rangle = i\int d^3x \langle B|\varphi^{in}(x)|A\rangle \overleftrightarrow{\frac{\partial}{\partial x_0}} f_{\mathbf{k}}$$

$$= \lim_{x_0 \to -\infty} i\int d^3x \langle B|\varphi(x)|A\rangle \overleftrightarrow{\frac{\partial}{\partial x_0}} f_{\mathbf{k}} .$$

Using the 4-dimensional Gauss theorem,

$$\langle B|A\pi_{\mathbf{k}}\rangle = \lim_{x_0 \to +\infty} i\int d^3x \langle B|\varphi(x)|A\rangle \overleftrightarrow{\frac{\partial}{\partial x_0}} f_{\mathbf{k}}(x)$$

$$- i\int d^4x \frac{\partial}{\partial x_0}\left[\langle B|\varphi(x)|A\rangle \overleftrightarrow{\frac{\partial}{\partial x_0}} f_{\mathbf{k}}\right] .$$

The first term, written as $\langle B|(a_{\mathbf{k}}^{OUT})^{+}|A\rangle = \langle B\text{"minus"}\pi_{\mathbf{k}}|A\rangle$, is zero except for forward elastic scattering, so we ignore it from now on (it corresponds to "1" in the S -matrix expansion).

Meanwhile, derive the following useful formula:

$$\int d^4x \frac{\partial}{\partial x_0}\left[A(x) \overleftrightarrow{\frac{\partial}{\partial x_0}} f_{\mathbf{k}}(x)\right] = \int d^4x \left(A \frac{\partial^2}{\partial x_0^2} f_{\mathbf{k}} - \frac{\partial^2 A}{\partial x_0^2} f_{\mathbf{k}}\right)$$

$$= \int d^4x\, f_{\mathbf{k}} (\Box - m^2) A ,$$

where we have used the K-G equation and integrated twice by parts with $f_{\mathbf{k}}(x)$ vanishing on the boundary. Similarly,

$$\int d^4x \frac{\partial}{\partial x_0}\left[f_{\mathbf{k}}^{*} \overleftrightarrow{\frac{\partial}{\partial x_0}} A\right] = -\int d^4x\, f_{\mathbf{k}}^{*} (\Box - m^2) A .$$

With these we get the matrix element for pion absorption,

$$\langle B|A\pi_{\mathbf{k}}\rangle = i\int d^4x \frac{e^{ik\cdot x}}{\sqrt{2\omega V}} (-\Box + m^2) \langle B|\varphi(x)|A\rangle$$

and similarly for pion emission.

$$\langle B\pi_{\mathbf{k}}|A\rangle = i\int d^4x \frac{e^{-ik\cdot x}}{\sqrt{2\omega V}} (-\Box + m^2) \langle B|\varphi(x)|A\rangle$$

So in these, the simplest of the reduction formulae, we see that the $\pi_{\mathbf{k}}$ is effectively "reduced" or "contracted" out of the state vector.

For pion scattering, first reduce the incident pion.

$$\pi_{\mathbf{k}} + A \longrightarrow \pi_{\mathbf{k}'} + B$$

$$\langle B\pi_{\mathbf{k}'}|A\pi_{\mathbf{k}}\rangle = i\int d^4x \, \frac{e^{ik\cdot x}}{\sqrt{2\omega V}}(-\Box + m^2)\langle B\pi_{\mathbf{k}'}|\varphi(x)|A\rangle$$

$$\langle B\pi_{\mathbf{k}'}|\varphi(x)|A\rangle = \langle B|\, a^{out}_{\mathbf{k}'}\, \varphi(x)|A\rangle$$

$$= i\int d^3x' f^*_{\mathbf{k}'}(x') \overleftrightarrow{\frac{\partial}{\partial x_0'}} \langle B|\varphi^{out}(x')\varphi(x)|A\rangle$$

$$= \lim_{x_0' \to \infty} i\int d^3x' f^*_{\mathbf{k}'} \overleftrightarrow{\frac{\partial}{\partial x_0'}} \langle B|T(\varphi(x')\varphi(x))|A\rangle$$

The time ordering does no harm since x_0' is necessarily later than x_c. By Gauss' theorem this becomes

$$\langle B\pi_{\mathbf{k}'}|\varphi(x)|A\rangle = \lim_{x_0' \to -\infty} i\int d^3x' f^*_{\mathbf{k}}(x') \overleftrightarrow{\frac{\partial}{\partial x_0'}} \langle B|\varphi(x)\varphi(x')|A\rangle$$

$$+ i\int d^4x \frac{\partial}{\partial x_0'}\left[f^*_{\mathbf{k}'} \overleftrightarrow{\frac{\partial}{\partial x_0}} \langle B|T(\varphi(x')\varphi(x))|A\rangle\right].$$

The first term, written as $\langle B|\varphi(x)\, a^{in}_{\mathbf{k}'}|A\rangle$ is zero unless A contains $\pi_{\mathbf{k}'}$.

So using the formula for $\int d^4x \frac{\partial}{\partial x_0}[f^*_{\mathbf{k}} \overleftrightarrow{\frac{\partial}{\partial x_0}} A(x)]$ we get

$$\langle B\pi_{\mathbf{k}'}|\varphi(x)|A\rangle = i\int d^4x' f^*_{\mathbf{k}'}(-\Box + m^2)\langle B|T(\varphi(x')\varphi(x))|A\rangle,$$

and therefore

$$\langle B\pi_{\mathbf{k}'}|A\pi_{\mathbf{k}}\rangle = i^2\int d^4x' \int d^4x \, \frac{e^{-ik'\cdot x'}}{\sqrt{2\omega' V}}\frac{e^{ik\cdot x}}{\sqrt{2\omega V}}(-\Box' + m^2)(-\Box + m^2)\langle B|T(\varphi(x')\varphi(x))|A\rangle.$$

This formula is good except when $\mathbf{k} = \mathbf{k}'$ and A = B since we have ignored terms that contribute only for elastic processes in the strictly forward direction.

To obtain the reduction formula for $\langle B\pi_{\mathbf{k}'}\pi_{\mathbf{k}}|A\rangle$ simply let $\mathbf{k}\rightarrow -\mathbf{k}$.

The same technique can be used to reduce the matrix elements appearing in weak or e.m. processes.

$$\langle B\pi_{\mathbf{k}}|d(0)|A\rangle = i\int d^4x \frac{e^{-ik.x}}{\sqrt{2\omega V}}(-\Box + m^2)\langle B|T(\varphi(x)d(0))|A\rangle$$

where

$$d = j_{\mu,5\mu}^{1\pm i2,\,4\pm i5} \quad \text{or} \quad j_\mu^{em}\,.$$

Note the similarity between this and the formula for $\langle B\pi_{\mathbf{k}}|\varphi(x)|A\rangle$. A 1st order current interaction is very much like the emission or absorption of a "particle" - sometimes called a "spurion" - which can carry away internal quantum numbers such as isospin and strangeness.

The reduction formula for scattering is sometimes written in the form

$$\langle B\pi_{\mathbf{k}'}|A\pi_{\mathbf{k}}\rangle = i^2\int d^4x'\int d^4x \frac{e^{-ik'.x'}}{\sqrt{2\omega' V}}\frac{e^{ik.x}}{\sqrt{2\omega V}}\times$$

$$\times(-\Box'+m^2)(-\Box+m^2)\langle B|\Theta(x_0'-x_0)[\varphi(x'),\varphi(x)]|A\rangle.$$

This "retarded commutator" form is more convenient for proving dispersion relations since the integrand vanishes except in the forward light cone of $(x'-x)$. The two expressions are completely identical for physical values of $\mathbf{k}$. Their analytic structures are different, however, for complex values of k_0.

Finally we may just mention that, by using the reduction technique to express the S-matrix element as a T product, it is possible to derive the perturbation expansion of the S-matrix (in the Dyson form). In fact Low first obtained the reduction formula starting with Dyson's perturbation formula.

Single Soft-pion Emission (Absorption) in Strong Interaction Processes

We are interested in the emission process

$A \to B + \pi^\alpha$.

$$\langle \pi_q^\alpha B | A \rangle = i \int d^4x \frac{e^{-iq\cdot x}}{\sqrt{2\omega}} (-\Box + m_\pi^2) \langle B | \pi^\alpha(x) | A \rangle$$

$$= i \int d^4x \frac{e^{-iq\cdot x}}{\sqrt{2\omega}} (-\Box + m_\pi^2) \frac{\langle B | \partial_\mu j_{5\mu}^\alpha(x) | A \rangle}{C_\pi m_\pi^2}$$

$$= -\frac{q_\mu}{C_\pi m_\pi^2} \int d^4x \frac{e^{-iq\cdot x}}{\sqrt{2\omega}} (-\Box + m^2) \langle B | j_{5\mu}^\alpha(x) | A \rangle$$

(Although we have used the divergence-field identity, we have not yet made the usual approximations involved in many applications of PCAC. This is because $(-\Box + m_\pi^2)$ picks out only the π pole contribution in $\langle B | \partial_\mu j_{5\mu}^\alpha | A \rangle$.)

Consider now

$$\sqrt{2\omega} \langle \pi_q^\alpha B | A \rangle = -\frac{(q^2 + m_\pi^2)}{C_\pi m_\pi^2} q_\mu \int d^4x \, e^{-iq\cdot x} \langle B | j_{5\mu}^\alpha(x) | A \rangle .$$

On the mass shell [i.e. $\lim_{q^2 \to -m_\pi^2}$] with $q = p_A - p_B$, this expression agrees with the usual expression for $A \to B + \pi^\alpha$. However, this equation <u>defines</u> the amplitude for $A \to B + \pi^\alpha_{SOFT}$ even when the pion is off the mass shell. In particular, consider $q \to 0$ with m_π^2 fixed.

$$\lim_{q\to 0}\sqrt{2\omega}\,\langle \pi_q^\alpha B|A\rangle = \lim_{q\to 0}\left\{-\frac{q_\mu}{C_\pi}\int d^4x\, e^{-iq\cdot x}\, e^{i(p_A-p_B)\cdot x}\langle B|j^\alpha_{5\mu}(0)|A\rangle\right\}$$

$$= \lim_{q\to 0}\left\{-i\,\delta^{(4)}(p_B+q-p_A)\left(\frac{-iq_\mu}{C_\pi}\right)\langle B|j^\alpha_{5\mu}(0)|A\rangle\right\}$$

Note that energy-momentum conservation still holds. The LHS is usually finite since the amplitude for $A\to B + \pi^\alpha$ is expected to be nonvanishing even in the soft pion limit; but the RHS may, at first sight, appear to go to zero. Actually there is no contradiction, because $\langle B|j^\alpha_{5\mu}(0)|A\rangle$ can have a singularity going like $1/q$. A notable exception is when A and B are both single particle states, in which case $\langle B|j^\alpha_{5\mu}|A\rangle$ (e.g. β decay matrix element) has no $1/q$ singularity; this is not troublesome since the amplitude for $A\to B + \pi$ does go to zero as $q\to 0$ because $\bar{u}'\gamma_5 u\to 0$ etc.

To compute $\langle B\pi^\alpha|A\rangle$ in the soft pion limit, it suffices to consider only that part of $\langle B|j^\alpha_{5\mu}|A\rangle$ having a $1/q$ singularity. This is because nonsingular parts of $\langle B|j^\alpha_{5\mu}|A\rangle$ don't contribute when multiplied by q_μ. In terms of Feynman graphs, $\langle B|j^\alpha_{5\mu}|A\rangle$ represents all possible diagrams in which A goes to B under the action of a single axial interaction. So we ask which diagrams with a single axial insertion go like $1/q$.

(a) Axial interaction inserted in an internal line.

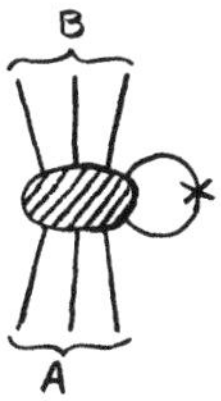

The diagram with insertion differs by an extra internal propagator. From specific examples it is not difficult to see that the additional propagator cannot give rise to a $1/q$ singularity. As a simple case consider one of the internal propagators.

$$\left(\frac{1}{p_{int}^2 + m^2 - i\varepsilon}\right) \xrightarrow{\text{insertion}} \left(\frac{1}{p_{int}^2 + m^2 - i\varepsilon}\right) \lim_{q\to 0} \frac{1}{(p_{int} - q)^2 + m^2 - i\varepsilon}$$

Recall now that there is a four-dimensional integration over p_{int} to be performed. We then see that this diagram does not exhibit singular behavior as $q \to 0$.

(b) External pion line terminated by an axial interaction.

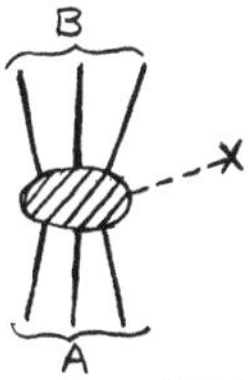

This diagram is given by $\langle B | J^{(\pi),\alpha} | A \rangle G_\pi q_\mu / (q^2 + m_\pi^2)$ which exhibits no singular behavior as $q \to 0$. (Note $m_\pi^2 \neq 0$ in this formalism.)

(c) <u>Axial interaction attached to an unterminated external line.</u>

(Axial insertion in an unterminated external pion line is forbidden by G-parity.) Take the nucleon as an example. The final nucleon line becomes

$$\bar{u}'(p') \xrightarrow{\text{insertion}} i\,\bar{u}'(p')\gamma_\mu\gamma_5\frac{\tau_a}{2}\frac{(-i\gamma\cdot p'+m_N)}{(p'+q)^2+m_N^2}F_A(0)$$

$$= F_A(0)\,\bar{u}'(p')\gamma_\mu\gamma_5\frac{\tau_a}{2}\left(\frac{\gamma\cdot p'+im_N}{2p'\cdot q}\right).$$

This does have a $1/q$ singularity. Similarly for the initial nucleon line.

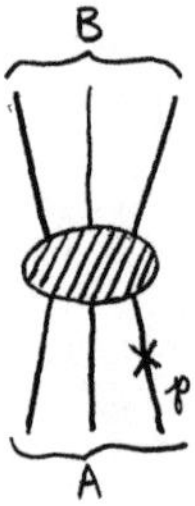

$$u(p) \xrightarrow{\text{insertion}} F_A(0)\left(\frac{\gamma\cdot p+im_N}{-2p\cdot q}\right)\gamma_\mu\gamma_5\frac{\tau_a}{2}u(p)$$

(d) For <u>an unterminated external line with different masses</u>, e.g. Λ and Σ^0,

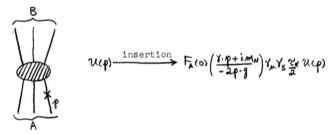

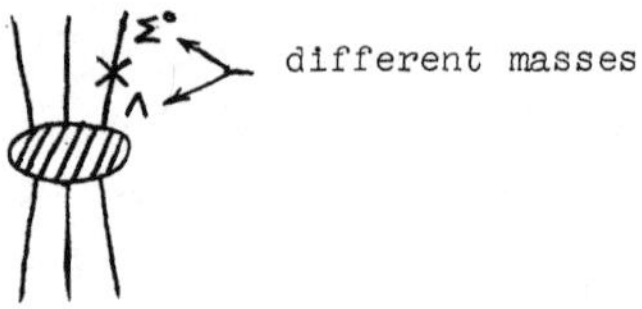

the denominator of the new propagator

$$\frac{1}{(p'-q)^2 + m_\Lambda^2} = \frac{1}{-m_\Sigma^2 + m_\Lambda^2 - 2p'\cdot q + q^2} \xrightarrow[q\to 0]{} \frac{1}{m_\Lambda^2 - m_\Sigma^2}$$

has no $1/q$ singularity. (We ignore the p-n mass difference since this would break charge independence, which we want to preserve here.)

To summarize,

$$\lim_{q\to 0} \langle B | j^\alpha_{5\mu}(0) | A \rangle_{p_B + q = p_A}$$

has $1/q$ singularities from (**c**)-type diagrams only. Recall that for $\langle \pi^\alpha_{q=0} B | A \rangle$ we are concerned with $(-iq_\mu/c_\pi)$ times this quantity; therefore to get the matrix element for soft-π emission,

$$\bar{u}'(p') \longrightarrow -i\left(\frac{F_A(0)}{2c_\pi}\right)\bar{u}'(p')\,\gamma\cdot q\,\gamma_5\,\tau_\alpha\left(\frac{\gamma\cdot p' + i m_N}{2p'\cdot q}\right)$$

$$= -i\left(\frac{G_{\pi NN} K_{\pi NN}(0)}{2m_N}\right)\bar{u}'(p')\,\gamma\cdot q\,\gamma_5\,\tau_\alpha\left(\frac{\gamma\cdot p' + i m_N}{2p'\cdot q}\right),$$

and similarly for the initial nucleon line. But this prescription is the same as inserting a soft-pion to external lines using the gradient coupling theory. So to compute A → B + π_{SOFT} when A → B is known, simply attach a soft pion according to the gradient coupling prescription. More precisely, if

A = one nucleon (p) + any number of "hard" mesons
B = " " (p') + " " " " ",

then the transition matrix element for A → B + π_{soft} is given by

$$-i\left(\frac{G_{\pi NN} K_{\pi NN}(0)}{2m_N}\right)\bar{u}'(p')\left\{\gamma\cdot q\,\gamma_5\,\tau_\alpha\left[\frac{\gamma\cdot p' + i m_N}{2p'\cdot q}\right]\mathcal{M} + \mathcal{M}\left[\frac{\gamma\cdot p + i m_N}{-2p\cdot q}\right]\gamma\cdot q\,\gamma_5\,\tau_\alpha\right\}u(p),$$

where $\bar{u}'(p)\, \mathcal{M}\, u(p)$ is the transition matrix element for $A \to B$ (directly measurable since here all particles are on the mass shell).

The soft pion formula was first derived by Nambu and Lurié ('62) and by Nambu and Schrauner ('62) using chirality (Q_5^α) conservation with $m_\pi = 0$. Our approach follows that of Adler ('65), with $m_\pi \neq 0$ but $q_\pi \to 0$. The two approaches lead to identical results since in Adler's approach the m_π^2 dependence drops out in the $q \to 0$ limit.

The soft-pion process is analogous to the infrared process in QED. As is well known,

$$e^- + Z \to e^- + Z \quad \text{(Mott scattering)}$$

is related to

$$e^- + Z \longrightarrow e^- + Z + \gamma_{\text{soft}} \quad \text{(bremsstrahlung)}$$

in the soft-photon limit. Again, only those diagrams in which the photon lines are attached to external lines contribute. But there is an important difference. In the soft-pion case the numerator is proportional to q, due to the gradient coupling ($\gamma \cdot q \gamma_5 \tau_\alpha$). So we have $\sim q/p \cdot q$, and the soft-pion matrix element has no infrared divergence. In contrast, the numerator in the soft-photon case does not go like k because of the γ_μ vertex. As a result, the bremsstrahlung matrix element diverges as $k \to 0$. This is the well-known infrared catastrophe.

Adler Consistency Condition

The soft-pion technique can be used to relate the πNN vertex to off-mass-shell π-N scattering. To conform with Adler's ('65) paper, consider the case where the absorbed π is soft and the emitted π is on the mass shell. Specifically we want to show that

$$\pi_{soft} + A \longrightarrow B \text{ is related to}$$

$$(A = \text{nucleon}) \xrightarrow[\text{vertex}]{\pi NN} (B = \text{nucleon} + \pi_{\text{mass shell}}).$$

The πNN vertex is characterized by $i\, G_{\pi NN}\, \gamma_5 \tau_\beta$; so for the outgoing π (nonsoft) use γ_5 coupling and for the incident π (soft) use gradient coupling.

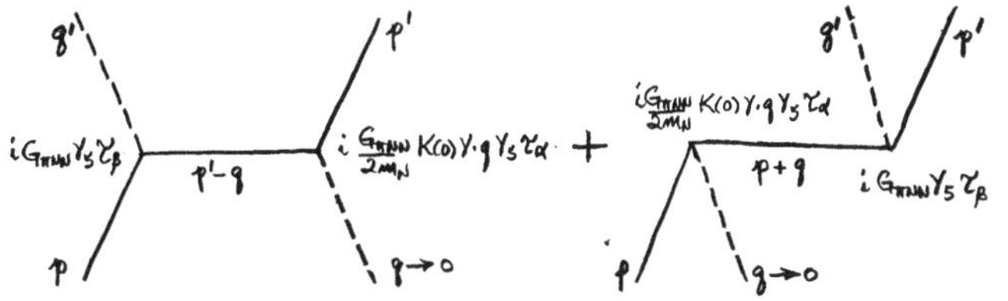

Thus the covariant transition matrix for off-mass-shell π-N scattering is

$$\mathcal{M} = i^2 \frac{G_{\pi NN}^2}{2m_N} K_{\pi NN}(0)\, \bar{u}' \left\{ \gamma\cdot q\, \gamma_5 \tau_\alpha \left(\frac{-i\gamma\cdot p' + m_N}{-2p'\cdot q} \right) i\gamma_5 \tau_\beta + i\gamma_5 \tau_\beta \left(-\frac{-i\gamma\cdot p + m_N}{2p\cdot q} \right) \gamma\cdot q\, \gamma_5 \tau_\alpha \right\} u$$

$$= \bar{u}' \left\{ -\frac{G_{\pi NN}^2}{m_N} K_{\pi NN}(0)\, \delta_{\alpha\beta} + \frac{G_{\pi NN}^2}{2m_N} K_{\pi NN}(0)\, i\gamma\cdot q \left[\frac{m_N}{p'\cdot q} \tau_\alpha \tau_\beta + \frac{m_N}{p\cdot q} \tau_\beta \tau_\alpha \right] \right\} u.$$

Quite generally $\mathcal{M}$ can be expressed (in the usual CGLN way) as

$$\mathcal{M} = \bar{u}' \left\{ (-A^{(+)} + i\gamma\cdot Q B^{(+)})\, \delta_{\alpha\beta} + (-A^{(-)} + i\gamma\cdot Q B^{(-)}) \frac{[\tau_\alpha, \tau_\beta]}{2} \right\} u.$$

where $Q \equiv (q + q')/2$. This decomposition is still possible for our off-mass-shell amplitude as long as the nucleons are on the mass shell, i.e.

$$(i\gamma \cdot p + m_N)u = 0 , \qquad \bar{u}'(i\gamma \cdot p' + m_N) = 0 .$$

Moreover, because of energy-momentum conservation $(p+q=p'+q')$,

$$i\bar{u}'\gamma \cdot Q u = i\bar{u}'\gamma \cdot \frac{(q+p-p'+q)}{2} u = i\bar{u}'\gamma \cdot q u .$$

When all four particles are on the mass shell, $A^{(+)}$ and $B^{(\pm)}$ are functions of two invariant scalars, e.g. s and t. But in our case, one of the particles is off the mass shell; so we need 3 scalar variables to specify the amplitude, namely (Adler's choice):

(1) $\nu = -\frac{P \cdot Q}{m_N} = -\frac{(p+p')\cdot(q+q')}{4m_N} = -\frac{(p+p')\cdot q}{2m_N}$

(since $q' = p - p' + q$ and $(p+p')\cdot(p-p') = 0$)

(2) $\nu_B = \frac{q' \cdot q}{2m_N}$ $\quad \left(= \frac{(p-p')\cdot q}{2m_N}\right.$ as $\left. q \to 0\right)$

(3) q^2 ($\neq -m_\pi^2$ when the incident π is off mass shell).

In terms of these new variables we can write the matrix element for $\pi_{\text{soft}} + N \to \pi_{\text{mass shell}} + N$ as

$$\mathcal{M} = \bar{u}'\left\{-\frac{G_{\pi NN}^2}{m_N} K_{\pi NN}^{(0)} \delta_{\alpha\beta} - i\gamma \cdot q \frac{G_{\pi NN}^2}{2m_N} K_{\pi NN}^{(0)} \left(\frac{\tau_\alpha \tau_\beta}{\nu_B + \nu} - \frac{\tau_\beta \tau_\alpha}{\nu_B - \nu}\right)\right\} u .$$

Note that we are talking about the scattering amplitude at the unphysical point (0,0,0) in (ν, ν_B, q^2) space. Comparison with the CGLN decomposition gives Adler's "consistency condition" for π-N scattering:

$$A^{(+)}(\nu = 0, \nu_B = 0, q^2 = 0) = \frac{G_{\pi NN}^2}{m_N} K_{\pi NN}^{(0)} .$$

(It is wrong to say "self-consistency condition", since there is nothing self-consistent about it.)

To check this relation, Adler's original procedure was to assume

$$A^{(+)}(0,0,0) \approx A^{(+)}(0,0,-m_\pi^2)$$

and estimate $A^{(+)}(0,0,-m_\pi^2)$ using forward dispersion relations - whereupon good agreement was obtained. A somewhat simpler method, which at the same time emphasizes the physical significance of Adler's condition, is to evaluate $-A^{(+)} + i\gamma\cdot Q\, B^{(+)}$ at the threshold of physical (on-mass-shell) π-N scattering, assuming that the dependence of $A^{(+)}$ and $B^{(+)}$ on ν, ν_B is still given by the formula we derived at $\nu = \nu_B = 0$. At threshold we have $q = q' = (\vec{0}, im_\pi)$, and hence

$$(\nu, \nu_B, q^2) = \left(m_\pi, -\frac{m_\pi^2}{2m_N}, -m_\pi^2\right)$$
$$\bar{u}' i\gamma\cdot Q u = -m_\pi u^\dagger u .$$

According to our assumption the amplitudes $A^{(+)}$ and $B^{(+)}$ still take the form

$$A^{(+)} = \frac{G_{\pi NN}^2}{m_N} K_{\pi NN}(0)$$
$$B^{(+)} = \frac{G_{\pi NN}^2}{m_N} K_{\pi NN}(0) \frac{\nu}{\nu_B^2 - \nu^2}$$

even at $\left(m_\pi, -\frac{m_\pi^2}{2m_N}, -m_\pi^2\right)$.

This means that the isospin symmetric part of the full amplitude at threshold is given by

$$\mathcal{M}^{(+)} = \bar{u}'(-A^{(+)} + i\gamma\cdot Q B^{(+)})u$$
$$= u^\dagger\left[-\frac{G_{\pi NN}^2}{m_N} K_{\pi NN}^{(0)} - m_\pi \frac{G_{\pi NN}^2}{m_N} K_{\pi NN}^{(0)} \frac{m_\pi}{\frac{1}{4}\left(\frac{m_\pi^2}{m_N}\right)^2 - m_\pi^2}\right] u$$

$$\mathcal{M}^{(+)} = \frac{G_{\pi NN}^2}{4m_N} K_{\pi NN}(0) \left(\frac{m_\pi}{m_N}\right)^2 = 0 + \mathcal{O}\left(\frac{m_\pi^2}{m_N^3}\right)$$

(Note the huge cancellation.) Since

$$\mathcal{M}^{(+)} = \frac{2\mathcal{M}^{(3/2)} + \mathcal{M}^{(1/2)}}{3}, \quad \mathcal{M}^{(-)} = \frac{\mathcal{M}^{(3/2)} - \mathcal{M}^{(1/2)}}{3},$$

the isospin symmetric part of the S-wave π-N scattering length is

$$2a_3 + a_1 = 0 + \frac{1}{m_N}\mathcal{O}\left(\frac{m_\pi^2}{m_N^2}\right),$$

which is in good agreement with the experimental value,

$$2a_3 + a_1 = (0.00 \pm 0.01)/m_\pi .$$

From this we see that Adler's consistency condition is satisfied very well — provided

$$A^{(+)}\left(m_\pi, -\frac{m_\pi^2}{2m_N}, -m_\pi^2\right) = A^{(+)}(0,0,0)$$

and the dependence of $B^{(+)}$ on ν and ν_B at threshold is still given by the expression we obtained in the $q \to 0$ limit.

To appreciate the significance of this truly remarkable result, just compute the conventional nucleon exchange contribution to the on-mass-shell π-N scattering amplitude using γ_5 coupling at both vertices. Then at threshold we get

$$\mathcal{M}^{(+)} = -\frac{G_{\pi NN}^2}{m_N}\bar{u}u$$

which is $4(m_N/m_\pi)^2 = 180$ times as large as the estimate we get from Adler's condition. That the quantity $a_1 + 2a_3$ is so fantastically small had been one of the greatest mysteries in pion physics since the early 50's. It is fair to say that this mystery has finally been solved (once we accept the "smoothness" hypothesis of the off-mass-shell amplitude). We'll say more about low-energy π-N scattering later, when we treat this problem in the limit where both the incoming and outgoing pions are soft.

For π-π scattering in which one of the pions is soft and the other three are on the mass shell, the scattering amplitude is given by

$$-i\frac{q_\mu}{C_\pi}\cdot$$

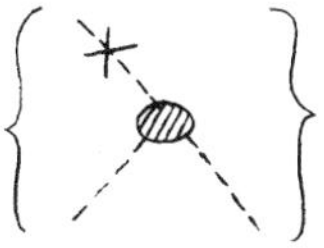

But this diagram is forbidden by G-parity because a $G = -1$ axial (spurion) iteraction cannot be attached to an external pion line. Thus we get Adler's consistency condition for π-π scattering: the amplitude vanishes if one of the pions is soft (and the other three pions on the mass shell).

Master Formula for Soft Pion Emission in Weak (e.m.) Processes

We now show that the amplitude for single-pion emission in weak (e.m.) processes can be computed exactly if the pion is soft. First recall that

$$\langle \pi^\alpha B | d(0) | A\rangle = i\int d^4x \, \frac{e^{-iq\cdot x}}{\sqrt{2\omega}}(-\Box + m_\pi^2)\langle B|T(\pi^\alpha(x)\, d(0))|A\rangle$$

is the amplitude for $A \xrightarrow[(e.m.)]{weak} B + \pi$ under the action of **d**.

In the soft pion limit we have

$$\lim_{q\to 0}\sqrt{2\omega}\,\langle \pi^\alpha B|d(0)|A\rangle = \frac{i}{C_\pi}\lim_{q\to 0}\left(\frac{q^2+m_\pi^2}{m_\pi^2}\right)\int d^4x\, e^{-iq\cdot x}\langle B|T(\partial_\mu j^\alpha_{5\mu}(x)\, d(0))|A\rangle.$$

Instead of this expression, it is more convenient to start with

$$q_\mu M_\mu = q_\mu \int d^4x\, e^{-iq\cdot x} \langle B|T(j^\alpha_{5\mu}(x)\, d(0))|A\rangle .$$

Write $q_\mu e^{-iq\cdot x} = i\partial_\mu e^{-iq\cdot x}$ and integrate by parts.

$$q_\mu M_\mu = -i\int d^4x\, e^{-iq\cdot x}\partial_\mu \langle B|T(j^\alpha_{5\mu}(x)\, d(0))|A\rangle$$

From

$$T(A(x)\,B(0)) = A(x)\,B(0)\,\theta(x_0) + B(0)\,A(x)\,\theta(-x_0)$$

and

$$\partial_\mu \theta(x_0) = \frac{1}{i}\frac{\partial}{\partial x_0}\theta(x_0)\,\delta_{\mu 4} = -i\,\delta(x_0)\,\delta_{\mu 4}$$

we get

$$q_\mu M_\mu = -i\int d^4x\, e^{-iq\cdot x}\{\langle B|T(\partial_\mu j^\alpha_{5\mu}(x)\, d(0))|A\rangle$$
$$-\,i\,\delta(x_0)\langle B|[j^\alpha_{54}(x), d(0)]|A\rangle\}.$$

In the $q\to 0$ limit, the 1st term is essentially the matrix element for $A \to B + \pi^\alpha_{soft}$ and the 2nd term is $\langle B|[Q^\alpha_5(0), d(0)]|A\rangle$. Thus we have the "master formula",

$$\lim_{q\to 0} \sqrt{2\omega}\langle \pi^\alpha B|d(0)|A\rangle = -\frac{i}{C_\pi}\langle B|[Q^\alpha_5(0), d(0)]|A\rangle$$
$$-\lim_{q\to 0}\frac{q_\mu}{C_\pi}\int d^4x\, e^{-iq\cdot x}\langle B|T(j^\alpha_{5\mu}(x)\, d(0))|A\rangle .$$

The last term, called the "pole term", is usually zero. Even when it's not zero, it can be calculated exactly - as we shall see later.

The master formula, essentially says that pion emission is due to that part of the interaction Lagrangian which disturbs the conservation of axial charge. A formula of this kind was first obtained by Nambu and

Schrauner ('62), but the derivation given here is due to Callan and Treiman ('66).

Using this formula, a variety of problems can be attacked, as we shall see in the following sections.

Leptonic Kaon Decay

(Callan and Treiman '66; Mathur, Okubo and Pandit '66)

Armed with the master formula, we derive relations among the amplitudes for

$$\left.\begin{array}{lll} K_{\mu 2} & : & K^+ \to \mu^+ + \nu' \\ K_{\mu 3}(K_{e3}) & : & K^+ \to \pi^0 + \mu^+ + \nu' \\ K_{e4} & : & K^+ \to \pi^+ + \pi^- + e^+ + \nu \end{array}\right\} \Delta S = \Delta Q = 1$$

with

$$d(0) = j^{4-i5}_{\mu, 5\mu}(0) .$$

For π^0 emission; $\alpha = 3$

$$[Q_5^3, j_\mu^{4-i5}] = -\tfrac{1}{2} j_{5\mu}^{4-i5} \qquad (f_{345} = -f_{354} = \tfrac{1}{2}),$$

$$[Q_5^3, j_{5\mu}^{4-i5}] = -\tfrac{1}{2} j_\mu^{4-i5} .$$

For π^+ emission; $\alpha = (1-i2)/\sqrt{2}$

$$[Q_5^{1-i2}, j^{4-i5}_{\mu,5\mu}] = 0 \qquad (f_{147} = f_{246} = f_{257} = -f_{156} = \tfrac{1}{2}).$$

Another (more elegant) way to see this is that Q^{4-i5} followed by Q^{1-i2} takes K^+ into π^-.

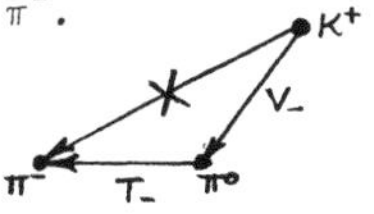

But no octet generator does this. Since the commutation relations among octet generators are supposed to be closed, the commutator must be zero.

For π^- emission; $\alpha = (1+i2)/\sqrt{2}$

$$[Q_5^{1+i2}, j_{\mu,5\mu}^{4-i5}] = - j_{5\mu,\mu}^{6-i7}.$$

Apply the master formula to $K_p^+ \longrightarrow \pi_q^0 + \mu^+ + \nu$!
Since this is a $J^P = 0^- \rightarrow 0^-$ transition, only the vector current contributes. By the master formula

$$\lim_{q\to 0} \sqrt{2q_0} \langle \pi^0 | j_\mu^{4-i5}(0) | K^+ \rangle = -\frac{i}{C_\pi} \langle 0 | [Q_5^3(0), j_\mu^{4-i5}(0)] | K^+ \rangle$$

$$= \frac{i}{2C_\pi} \langle 0 | j_{5\mu}^{4-i5}(0) | K^+ \rangle,$$

where we have ignored the pole term which can be shown to vanish in this case. (We'll say more about the pole terms later.) So CA + PCAC relate the off-mass-shell $K_{\mu 3}$ amplitude to the $K_{\mu 2}$ amplitude. The most general form for the $K_{\mu 3}$ matrix element is

$$\sqrt{4p_0 q_0} \langle \pi^0 | j_\mu^{4-i5}(0) | K^+ \rangle = -\frac{1}{\sqrt{2}} \{ (p+q)_\mu f_+ + (p-q)_\mu f_- \}$$

where the $(-1/\sqrt{2})$ factor is an F-type SU(3) C-G coefficient which normalizes the matrix element so that in the exact SU(3) limit $f_+ = 1$ at $(p-q)^2 = 0$. Note that when π^0 is off the mass shell, f_+ and f_- depend on both $(p-q)^2$ and q^2, whereas on the mass shell, they are functions of $(p-q)^2$ only. For $K_{\mu 2}$

$$\sqrt{2p_0} \langle 0 | j_{5\mu}^{4-i5}(0) | K^+ \rangle = i \sqrt{2}\, C_\pi p_\mu,$$

and therefore

$$\lim_{q\to 0} [(p+q)_\mu f_+ + (p-q)_\mu f_-] = \frac{C_K}{C_\pi} p_\mu$$

$$[f_+ + f_-]_{\substack{(p-q)^2 = -m_K^2 \\ q^2 = 0}} = \frac{C_K}{C_\pi} \approx 1.3 .$$

Note that on the mass shell

$$f_+ = 1 + \mathcal{O}(\lambda_{MS}^2) \quad \text{at} \quad (p-q)^2 = 0 \,, \quad q^2 = -m_\pi^2$$

by the Ademollo-Gatto theorem, where the coupling parameter λ_{MS} characterizes the SU(3) breaking. In the exact SU(3) limit ($\lambda_{MS} \longrightarrow 0$),

$$\left\{ \begin{matrix} f_+ = 1 \\ f_- = 0 \end{matrix} \right\}, \quad C_K = C_\pi \quad \text{at} \quad (p-q)^2 = 0 \,, \quad q^2 = -m_\pi^2 .$$

So our CA result gives some information on how SU(3) is broken. The CA prediction can in principle be tested - provided we assume the usual small q^2 dependence of $f_\pm$ between $q^2 = -m_\pi^2$ and 0. But comparison with experiment is difficult since f_-/f_+ and the $(p-q)^2$ dependence of $f_\pm$ is not well known. (Various experiments are not in agreement with each other.)

Now look at

$$K_{e4}: \quad K_p^+ \longrightarrow \pi_{q^+}^+ + \pi_{q^-}^- + e^+ + \nu .$$

This is a predominantly axial decay since the vector part which is of the form $\varepsilon_{\mu\nu\lambda\sigma}\, p_\nu q_\lambda^+ q_\sigma^-$ leads to a very small contribution (essentially because the Q value is not very large). So we apply the master formula to the axial vector part.

$$\lim_{q^+ \to 0} \sqrt{2q_0^+} \langle \pi^+\pi^- | j_{5\mu}^{4-i5} | K^+ \rangle = 0$$

$$\lim_{q^- \to 0} \sqrt{2q_0^-} \langle \pi^+\pi^- | j_{5\mu}^{4-i5} | K^+ \rangle = \frac{i}{\sqrt{2}\, C_\pi} \langle \pi^+ | j_\mu^{6-i7} | K^+ \rangle = \frac{i}{C_\pi} \langle \pi^0 | j_\mu^{4-i5} | K^+ \rangle ,$$

where in the last step the matrix elements are related by an SU(3) C-G coefficient.

Thus K_{e4} is related to $K_{\mu3}$, just as $K_{\mu3}$ is related to $K_{\mu2}$.

The most general form of the K_{e4} matrix element is

$$\sqrt{8p_0 q_0^+ q_0^-}\,\langle\pi^+\pi^-|\,j_{5\mu}^{4-i5}\,|K^+\rangle =$$

$$= (q^+ + q^-)_\mu f_1 + (q^+ - q^-)_\mu f_2 + (p - q^+ - q^-)_\mu f_3 .$$

On the mass shell the f's are functions of 3 variables:

$$p\cdot q^+ ,\ p\cdot q^- ,\ (q^+ + q^-)^2.$$

Off the mass shell the f's are functions of 4 variables:

$$p\cdot q^+ ,\ p\cdot q^- ,\ (q^+ + q^-)^2, \text{and } (q^+)^2 \text{ or } (q^-)^2.$$

[The last term , $(-p^{(e)} + p^{(\nu)})_\mu f_3$, when dotted with the lepton current gives a negligible contribution $(\sim m_e)$.] So

$$q^+ \to 0 : \begin{cases} f_1 = f_2 \\ f_3 = 0 \end{cases}$$

$$q^- \to 0 : \begin{cases} |f_1 + f_2| = |f_+ - f_-|/\sqrt{2}\,C_\pi \\ |f_3| = |f_+ + f_-|/\sqrt{2}\,C_\pi . \end{cases}$$

The rapid variation of f_3 can be shown to be not at all disturbing if the K pole is properly taken into account (Weinberg '66). But in any case the f_3 term gives a negligible contribution to the total decay rate. Experimentally

$$\frac{\langle f_2\rangle_{\text{average}}}{\langle f_1\rangle_{\text{average}}} = 0.9 \pm 0.2 ,$$

which is consistent with $f_1 = f_2$;

and the absolute K_{e4} rate agrees with the CA prediction to within 20 ± 20%.

Photoproduction

$$\gamma_k + N_p \longrightarrow \pi^\alpha_{q\,\text{soft}} + N'_{p'}$$

$$\lim_{q\to 0} \sqrt{2q_0}\,\langle \pi^\alpha N' | j_\mu^{(em)} | N\rangle = \frac{-i}{C_\pi}\langle N' | [Q_5^\alpha, j_\mu^{(em)}] | N\rangle + \text{"pole"}$$

$$= \frac{1}{C_\pi}\,\varepsilon_{\alpha 3\beta}\langle N' | j_{5\mu}^\beta | N\rangle + \text{"pole"}$$

Only the j_μ^3 part of $j_\mu^{(em)}$ contributes, because j_μ^8 commutes with $Q_5^{1,2,3}$. As $q \to 0$, we have $p' \to k + p$. But for a real photon, energy-momentum conservation demands that also $k \to 0$. Therefore ($p' - p \to 0$)

$$\langle N' | j_{5\mu}^\beta | N\rangle \longrightarrow \frac{m_N}{\sqrt{EE'}}\,F_A(0)\,\bar{u}' i\gamma_\mu\gamma_5 \frac{\tau_\beta}{2} u\,,$$

and the photoproduction matrix element becomes (ε_μ= photon polarization)

$$e\varepsilon_\mu \frac{F_A(0)}{C_\pi}\varepsilon_{\alpha 3\beta}\,\bar{u}' i\gamma_\mu\gamma_5 \frac{\tau_\beta}{2} u \overset{\text{G-T.}}{=} e\varepsilon_\mu \frac{G_{\pi NN}}{2m_N}\varepsilon_{\alpha 3\beta}\,\bar{u}' i\gamma_\mu\gamma_5 \tau_\beta u\,.$$

This is the Kroll-Ruderman theorem which states that in the $q \to 0$, $k \to 0$ limit the photoproduction matrix element is given by

$$e\frac{G_{\pi NN}}{2m_N}\,\vec{\varepsilon}\cdot\langle\vec{\sigma}\rangle \begin{Bmatrix} \sqrt{2} \\ 0 \\ -\sqrt{2} \end{Bmatrix} \quad \text{for} \quad \begin{Bmatrix} \gamma + p \to \pi^+ + n \\ \gamma + n \to \pi^- + p \\ \gamma + p \to \pi^0 + p \end{Bmatrix}.$$

Note that this matrix element can be obtained directly from the gradient coupling theory by making the replacement

$$\partial_\mu \pi^\pm \longrightarrow (\partial_\mu \mp i e A_\mu)\pi^\pm.$$

Until now we have ignored the pole term, which for photoproduction is

$$-\lim_{q\to 0} \frac{q_\nu}{C_\pi} \int d^4x\, e^{-iq\cdot x} \langle N'| T(j_{5\nu}^{\alpha}(x)\, j_\mu^{(em)}(0))|N\rangle .$$

This possible contribution can be calculated in two ways:

(1) Noncovariant method - Perform the time integration first.

$$\int \equiv \int d^4x\, e^{-iq\cdot x} \langle N'| T(j_{5\nu}^{\alpha}(x)\, j_\mu^{(em)}(0))|N\rangle$$

$$= \int d^3x\, e^{-i\vec{q}\cdot\vec{x}} \Big[\int_0^\infty dx_0\, e^{iq_0x_0} \langle N'| j_{5\nu}^{\alpha}(x)\, j_\mu^{(em)}(0)|N\rangle$$

$$+ \int_{-\infty}^0 dx_0\, e^{iq_0x_0} \langle N'| j_\mu^{(em)}(0)\, j_{5\nu}^{\alpha}(x)|N\rangle \Big]$$

Insert a complete set of states and use

$$\langle N'| j_{5\nu}^{\alpha}(x)|m\rangle = e^{-i(E_m-E')x_0} \langle N'| j_{5\nu}^{\alpha}(\vec{x},0)|m\rangle$$

$$\int_0^\infty dx_0\, e^{-iEx_0} = \frac{-i}{E-i\epsilon},$$

whereupon

$$\int = -i \int d^3x\, e^{-i\vec{q}\cdot\vec{x}} \sum_m \Big[\frac{\langle N'| j_{5\nu}^{\alpha}(\vec{x},0)|m\rangle\langle m| j_\mu^{(em)}(0)|N\rangle}{E_m - E' - q_0 - i\epsilon}$$

$$+ \frac{\langle N'| j_\mu^{(em)}(0)|m\rangle\langle m| j_{5\nu}^{\alpha}(\vec{x},0)|N\rangle}{E_m - E + q_0 - i\epsilon} \Big].$$

Multiply this by $-q_\nu / C_\pi$ and let $q \to 0$. The only intermediate state that survives is degenerate with the initial and final states, ($E_m = E' = E$), viz. the single nucleon state. This pole contribution, being proportional to

$$\frac{F_A(0)}{2C_\pi} = \frac{G_{\pi NN}}{2m_N},$$

is just what we would expect from

computed according to the noncovariant perturbation theory with the gradient coupling prescription.

(ii) Covariant method - Identify $\int$ with a complete set of Feynman diagrams in which a single axial interaction is attached to the $\gamma N\bar{N}$ vertex. As in the single soft π emission in strong processes, convince yourself that the only graphs that survive are those in which the axial interaction is attached to an external nucleon line; so only the nucleon pole contributes.

In our case, choose $\vec{k}$ along the z-axis so that $\mu = 1,2$ only.

$$\langle N' | j_{1,2}^{(em)} | N \rangle \sim \mathcal{O}(k)$$

If both k and $q \longrightarrow 0$, the pole term does not contribute. Only the Kroll-Ruderman term (due to the equal-time commutator) survives in this limit.

For electroproduction of soft π (first treated by Nambu and Schrauner), $q \to 0$ but $k^2 = (p' - p)^2 \neq 0$, and the equal-time commutator term now involves $F_A\,(-k^2 \neq 0)$. Also the pole contribution [which can be written in terms of $F_{1,2}^{(V,S)}(-k^2)$] cannot be neglected when $k \neq 0$. Altogether, it is possible to derive low energy theorems for 6 x 3 = 18 invariant amplitudes (The 6 comes from

Diracology and gauge invariance; the 3 from 2 for isovector photon + 1 for isoscalar photon.), which can be converted into dispersion-theoretic sum rules by expressing the low energy amplitudes as appropriate integrals over absorptive parts. Actually only three of them turn out to be nontrivial relations (Adler and Gilman; Fubini, Furlan and Rosetti; Furlan, Jengo and Remidi; Riazuddin and Lee '66).

The main results are:

(i) For the nucleon, $F_2^{(S)}(0) = (\mu_p + \mu_n)/2$ is predicted to be small - in agreement with observation.

(ii) $\langle r^2 \rangle_A \approx \frac{1}{2} \langle r^2 \rangle_1^{(V)}$, where

$$\langle r^2 \rangle_A = 6 \frac{dF_A^{(+)}}{dt} \quad , \quad \langle r^2 \rangle_1^{(V)} = 6 \frac{dF_1^{(V)}}{dt} .$$

Alternative Formalism Based on the Divergence Condition

Instead of working with $[Q_5, j^\beta_{\mu,5\mu}]$, we can work with the divergence condition, which in the quark model reads:

$$\partial_\mu j^\alpha_{5\mu} = i m \bar{q} \gamma_5 \lambda_\alpha q + O(\delta m) + f_{\alpha\beta\gamma} j^\beta_{5\mu} W^\gamma_\mu + f_{\alpha\beta\gamma} j^\beta_\mu W^\gamma_{5\mu} .$$

Comparing this with $\partial_\mu j^\alpha_{5\mu} = C_\pi m_\pi^2 \pi^\alpha$ when $W_{\mu,5\mu} = 0$, we infer a modification of PCAC in the presence of external fields as follows:

$$\partial_\mu j^\alpha_{5\mu} = C_\pi m_\pi^2 \pi^\alpha + f_{\alpha\beta\gamma} j^\beta_{5\mu} W^\gamma_\mu + f_{\alpha\beta\gamma} j^\beta_\mu W^\gamma_{5\mu} .$$

In the case of photoproduction

$$W^\gamma_\mu j^\gamma_\mu = e A_\mu \left(j^3_\mu + \frac{1}{\sqrt{3}} j^8_\mu \right) , \quad W^\gamma_{5\mu} = 0 ,$$

and therefore

$$\partial_\mu j^\alpha_{5\mu} = C_\pi m_\pi^2 \pi^\alpha + e\, \varepsilon_{\alpha\beta 3} j^\beta_{5\mu} A_\mu \qquad (\alpha = 1,2,3).$$

Now take the matrix element and integrate.

$$\frac{i}{C_\pi}\int d^4x\, e^{-iq\cdot x}\langle N'|\partial_\mu j^\alpha_{5\mu}|\gamma N\rangle =$$

$$= i\int d^4x\, e^{-iq\cdot x} m_\pi^2 \langle N'|\pi^\alpha|\gamma N\rangle + \frac{ie}{C_\pi}\int d^4x\, e^{-iq\cdot x}\varepsilon_{\alpha\beta 3}\langle N'|j^\beta_{5\mu}A_\mu|\gamma N\rangle$$

By means of reduction formula, the pion term is related to the soft π photoproduction amplitude. And since A_μ acts as a free field that annihilates a photon (to lowest order in e.m.), we can contract out the photon in the last term. So for $q\to 0$,

$$-\frac{q_\mu}{C_\pi}\int d^4x\, e^{-iq\cdot x}\langle N'|j^\alpha_{5\mu}|N\rangle =$$

$$= \sqrt{2q_0}\langle\pi^\alpha N'|\gamma N\rangle + i\frac{e}{C_\pi}\int d^4x\, e^{-iq\cdot x}\varepsilon_{\alpha\beta 3}\langle N'|j^\beta_{5\mu}|N\rangle\, \varepsilon_\mu \frac{e^{ik\cdot x}}{\sqrt{2k_0}},$$

and we get the result

$$\lim_{q\to 0}\sqrt{4q_0k_0}\langle\pi^\alpha N'|\gamma N\rangle = -i(2\pi)^4\delta^{(4)}(p+k-p'-q)\, e\varepsilon_\mu \frac{1}{C_\pi}\varepsilon_{\alpha\beta 3}\langle N'|j^\beta_{5\mu}(0)|N\rangle$$

$$-\lim_{q\to 0}\frac{q_\mu}{C_\pi}\int d^4x\, e^{-iq\cdot x}\sqrt{2k_0}\langle N'|j^\alpha_{5\mu}|\gamma N\rangle.$$

Thus the photoproduction matrix element as $q\to 0$ is again given by the Kroll-Ruderman term + the nucleon pole term - just as we got from the master formula and CA. If you don't like the CA approach, then you can rely on the divergence condition method - they are equivalent. Moreover, this equivalence can also be demonstrated for semileptonic processes treated earlier. This has been emphasized particularly by Veltman ('66). (See also Adler '65).

$K_{\pi 3}$ (τ) Decay

The current-current form of interaction is not well-established for nonleptonic weak processes since it is difficult to compute using $j_\mu^{1-i2}\ j_\mu^{4-i5}$, etc.. For pedagogical purposes it is somewhat simpler to assume that the basic interaction Lagrangian for nonleptonic processes is given by a scalar and a pseudoscalar density made up of quark fields. Assume CP conservation and the $T = 1/2$ rule.

Parity conserving: $\mathcal{L}^{(+),6} = g_{NL} \bar{q} \frac{\lambda_6}{2} q$ $\qquad C = P = +1$

Parity violating: $\mathcal{L}^{(-),7} = i g_{NL} \bar{q} \gamma_5 \frac{\lambda_7}{2} q$ $\qquad C = P = -1$

$CP = +1$ requires the presence of λ_6 for P conserving and λ_7 for P violating (See Chapter I).

τ decay is P conserving. So for the master formula we must compute $[Q_5^\alpha, \mathcal{L}^{(+),6}]$.

$$\{\gamma_5, \gamma_4\} = 0, \quad [\gamma_5, \gamma_4] = -2\gamma_4\gamma_5$$

Therefore the structure constant is of the d-type and multiplies the isospin transform of the parity violating interaction.

$$[Q_5^\alpha, \mathcal{L}^{(+),6}] = i\, d_{\alpha 6 \gamma} \mathcal{L}^{(-),\gamma}$$

In particular:

for π^0 emission $[Q_5^3, \mathcal{L}^{(+),6}] = -\frac{i}{2}\mathcal{L}^{(-),6}$ $\qquad (d_{366} = -\frac{1}{2})$,

for π^- emission $[Q_5^{1+i2}, \mathcal{L}^{(+),6}] = \frac{i}{2}\mathcal{L}^{(-),4+i5}$ $\qquad (d_{164} = d_{265} = \frac{1}{2})$.

Take $K_L \to \pi^+\pi^-\pi^0$ as an example. Recall that K_L with CP=-1 is the sixth component of the pseudoscalar meson octet ($K_L = \varphi_6$).

By the master formula

$$\lim_{q^0\to 0}\sqrt{2q_0^0}\,\langle\pi^+\pi^-\pi^0|\mathcal{L}^{(+),6}|K_L\rangle = \frac{-i}{C_\pi}\langle\pi^+\pi^-|[Q_5^3,\mathcal{L}^{(+),6}]|K_L\rangle$$

$$= \frac{-1}{2C_\pi}\langle\pi^+\pi^-|\mathcal{L}^{(-),6}|K_L\rangle .$$

This last term is the matrix element for the hypothetical CP violating interaction $K_L \to \pi^+\pi^-$ (not to be confused with the experimentally observed, CP violating $K_L \to \pi^+\pi^-$ interaction, which is much weaker). This sounds strange. However, we can rotate 90° about the 3rd axis in U-spin using the generator (orthogonal to $\lambda^{(em)}$)

$$\frac{1}{2}\left(-\frac{\lambda_3}{2}+\sqrt{3}\,\frac{\lambda_8}{2}\right).$$

The $\pi^+\pi^-$ system has $T_3 = Y = 0$ and is therefore unchanged. But a 90° rotation about the U_3 axis interchanges the 6th and 7th components, (apart from a possible minus sign) just as a 90° rotation about the T_3 axis interchanges the 1st and 2nd components. Hence

$$\lim_{q^0\to 0}\sqrt{2q_0^0}\,\langle\pi^+\pi^-\pi^0|\mathcal{L}^{(+),6}|K_L\rangle = \frac{1}{2C_\pi}\langle\pi^+\pi^-|\mathcal{L}^{(-),7}|K_S\rangle,$$

where we now have the matrix element for the CP conserving $K_S \to 2\pi$ ($K_S = Q^7$, CP = +1). This equation says that the off-mass-shell $K_{\pi 3}$ decay is related to the $K_{\pi 2}$ decay.

As another example look at the charged mode, $K^+ \to \pi^+\pi^+\pi^-$.

$$\lim_{q^-\to 0}\sqrt{2q_0^-}\,\langle\pi^+\pi^+\pi^-|\mathcal{L}^{(+),6}|K^+\rangle = \frac{1}{2\sqrt{2}\,C_\pi}\langle\pi^+\pi^+|\mathcal{L}^{(-),4+i5}|K^+\rangle$$

But $\langle\pi^+\pi^+|\mathcal{L}^{(-),\,4+i5}|K^+\rangle = 0$ since $\mathcal{L}^{(-),\,4+i5}$ <u>raises</u> the the strangeness. Consequently,

$$\lim_{q\to 0}\sqrt{2q_0^-}\,\langle\pi^+\pi^+\pi^-|\mathcal{L}^{(+),6}|K^+\rangle = 0 .$$

These limits were first worked out by Callan and Treiman. They were discouraged because experimentally the (on-mass-shell) $K^+_{\pi 3}$ decay has lots of events with $\vec{q}^- \approx 0$. Subsequently it was realized by many people (Hara and Nambu '66; Elias and Taylor '66; etc.) that the $q = 0$ requirements need not be satisfied at $\vec{q} = 0$, $q^2 = -m_\pi^2$. The invariant $K^+_{\pi 3}$ amplitude is a function of scalar products constructed out of p, q^+, $q^{+\prime}$, and q^-. Furthermore, the experimental (on-mass-shell) amplitude is well approximated by $\alpha + \beta\omega_-$, where $\omega_- = q_0^-$ in the K rest frame. This means that the experimental amplitude is a linear function of $p \cdot q^- = -m_K \omega_-$. (Another possible scalar invariant is $p \cdot (q^+ + q^{+\prime})$. But this is expressible in t erms of $p \cdot q^-$. Note that B-E statistics requires symmetry under $q^+ \leftrightarrow q^{+\prime}$). The off-mass-shell amplitude can, of course, depend on $(q^-)^2$ also, but assume slow variation between $(q^-)^2 = 0$ and $-m_\pi^2$. This suggests that the $q^- \to 0$ amplitude is best approximated by the on-mass-shell amplitude extrapolated to $\omega_- \to 0$. In other words, $q^- \to 0$ doesn't mean $\vec{q} \longrightarrow 0$ but rather $\omega_- \longrightarrow 0$. ($\omega_- = 0$ implies $\omega_+ = \omega_+' = m_K/2$. This kinematically resembles $K_{\pi 2}$ decay.)

The experimental $K^+_{\pi 3}$ amplitude as a function of ω_- looks as follows (Nefkens '66):

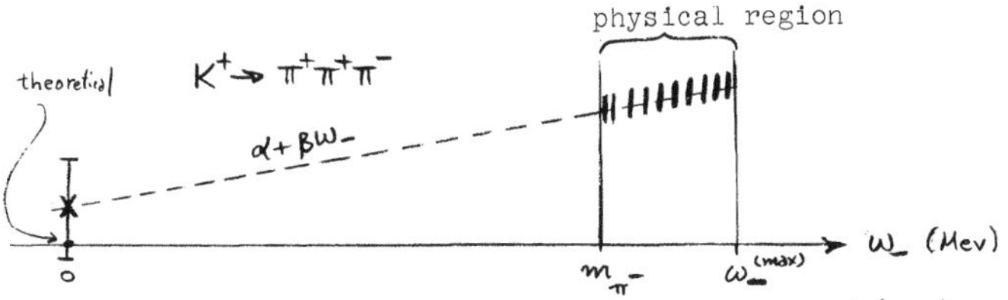

The extrapolated amplitude is consistent with zero ($\alpha \approx 0$) at $\omega_- = 0$. Similarly the $K_L \to \pi^+\pi^-\pi^0$ amplitude must be extrapolated to $\omega_0 = 0$ in order to see whether it agrees with that predicted from K_S decay.

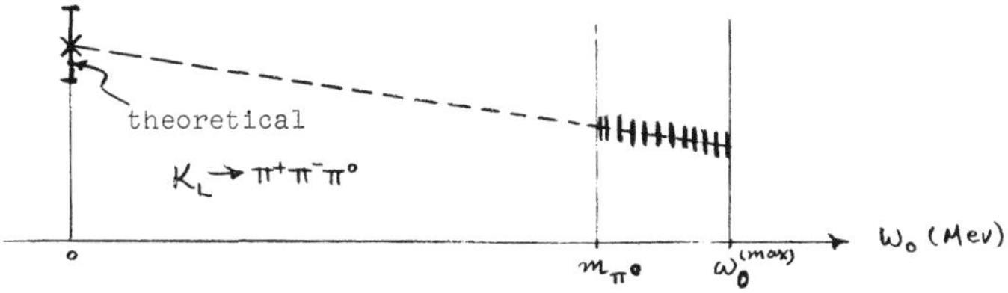

Again the theoretically predicted amplitude is consistent with the extrapolated value.

In terms of a Dalitz plot, CA and PCAC make predictions for 3 points, viz. when any of the 3 pions has zero energy.

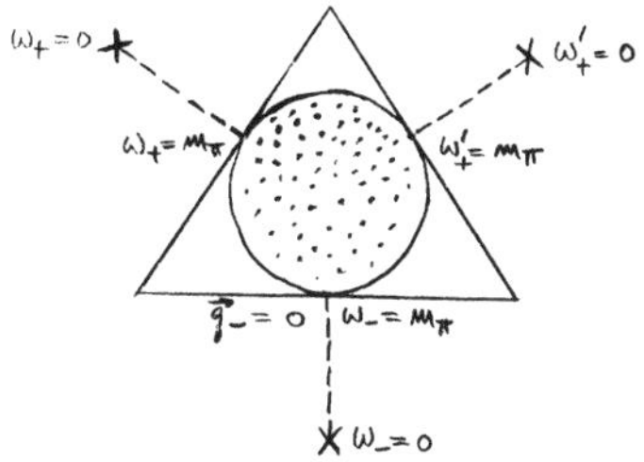

Prior to CA, people had to invoke final state interactions to explain this nonuniform Dalitz plot. But with the advent of CA, we can explain the spectrum slope without using strong π-π interactions. In fact, the success of the CA prediction would be hard to understand if there were strong π-π interactions!

S-wave (Parity Violating) Hyperon Decay

Again we work with interaction Lagrangians constructed from quark fields and take advantage of

$$[Q_5^\alpha , \mathcal{L}^{(-),7}] = -i\, d_{\alpha 7 \gamma}\, \mathcal{L}^{(+),\gamma} ,$$

$$\lim_{q \to 0} \sqrt{2q_0}\, \langle \pi^\alpha B^\epsilon | \mathcal{L}^{(-),7} | B^\delta \rangle = -\frac{1}{C_\pi} d_{\alpha 7 \gamma} \langle B^\epsilon | \mathcal{L}^{(+),\gamma} | B^\delta \rangle .$$

$\langle B^\epsilon | \mathcal{L}^{(+),\gamma} | B^\delta \rangle$ is the matrix element for the weak interaction transition mass, which by the SU(3) version of the Wigner-Eckart theorem can be written as

$$\langle B^\epsilon | \mathcal{L}^{(+),\gamma} | B^\delta \rangle \propto \frac{D}{F} d_{\epsilon\gamma\delta} + f_{\epsilon\gamma\delta} .$$

On the basis of the $\Delta T = 1/2$ rule alone, there are four independent S-wave (P violating) hyperson decay amplitudes:

1 Λ amplitude,

2 independent Σ amplitudes (2 ways of attaching $T=\frac{1}{2}$ to $T=1$ Σ),

1 Ξ amplitude.

In the CA approach the matrix element is characterized by two parameters, the overall strength and the D/F ratio. So we are left with 2 relations among the amplitudes, which can be shown to be:

$$(\text{i}) \quad A(\Sigma^+_+) = 0$$

$$(\text{ii}) \quad -2A(\Xi^-_-) = \sqrt{3}\, A(\Sigma^+_0) + A(\Lambda^0_-).$$

(Σ^+_0 etc. stands for $\Sigma^+ \to \pi^0 + p$). We could have guessed the first one since no octet generator connects Σ^+ with n; experimentally this relation is extremely good. The second one, called the Lee-Sugawara relation (first obtained in '64 using somewhat different considerations), is satisfied up to ≈ 10%. Further predictions can be made if we assume (universal spurion hypothesis)

$$\left(\frac{D}{F}\right)_{\substack{\text{Weak} \\ \text{transition mass}}} = \left(\frac{D}{F}\right)_{\substack{\text{medium strong} \\ \text{mass splitting}}},$$

so that now all S-wave amplitudes are expressible in terms of a single parameter, the overall strength (Hara, Nambu and Schechter '66). The following table compares the experimental with the predicted S-wave hyperon decay amplitudes where our normalization is such that

$$\Gamma_{S\,wave}(B \to B' + \pi^{\pm,0}) = \frac{|\vec{q}|}{8\pi m_B^2}\left[(m_B + m_{B'})^2 - m_\pi^2\right] \left|A(B_{\pm,0})\right|^2.$$

	Experiment ($\times 10^7$)	Theory ($\times 10^7$)
$\lvert A(\Lambda^0_-)\rvert$	3.4	3.4 (input)
$\lvert A(\Sigma^-_-)\rvert$	4.0	4.0
$\lvert A(\Sigma^+_+)\rvert$	0.0 ± 0.1	0
$\lvert A(\Sigma^+_0)\rvert$	(3.4 ± 0.3 , 2.5 ± 0.4)	2.8
$\lvert A(\Xi^-_-)\rvert$	4.4	3.9

We now discuss briefly how the above results for τ decay and S-wave hyperon decay get modified if we assume that the basic nonleptonic Lagrangian is of the current-current type. Within the framework of the current-current theory there have been two schools of thought concerning the empirically established $\Delta T = 1/2$ rule.

Model (i): The basic Lagrangian transforms like an octet and therefore satisfies the $\Delta T = 1/2$ rule. With CP invariance this necessarily implies that the Lagrangian is of the form

$$d_{6\alpha\beta}\,(j_\mu^\alpha + j_{5\mu}^\alpha)(j_\mu^\beta + j_{5\mu}^\beta)$$

which, if explicitly written out, reads

$$\frac{1}{2}\Big[(j_\mu^{4-i5} + j_{5\mu}^{4-i5})(j_\mu^{1+i2} + j_{5\mu}^{1+i2}) + (j_\mu^{1-i2} + j_{5\mu}^{1-i2})(j_\mu^{4+i5} + j_{5\mu}^{4+i5})$$
$$-2(j_\mu^6 + j_{5\mu}^6)(j_\mu^3 + j_{5\mu}^3 + \frac{1}{\sqrt{3}}j_\mu^8 + \frac{1}{\sqrt{3}}j_{5\mu}^8)\Big].$$

Note that both the charged (α, β = 1,2,4,5) and the neutral (α, β = 3,6,8) currents participate. (This is the price we must pay if we want the $\Delta T = 1/2$ rule in the basic interaction.)

Model (ii). The basic Lagrangian involves only those currents which appear in the semileptonic interaction. This means that the interaction is of the form

$$(j_\mu^{4-i5} + j_{5\mu}^{4-i5})(j_\mu^{1+i2} + j_{5\mu}^{1+i2}) + \text{H.c.}.$$

This has both a $\Delta T = 1/2$ part and a $\Delta T = 3/2$ part (or in the SU(3) classification both "8" and "27").

$$\mathcal{L} = \mathcal{L}_{1/2} + \mathcal{L}_{3/2}$$

In this model the experimentally observed $\Delta T = 1/2$ rule presumably arises from the fact that the dynamics of strong interactions is so arranged that the effect of $\mathcal{L}_{1/2}$ is somehow enhanced (or the effect of $\mathcal{L}_{3/2}$ is somehow suppressed).

Let us now come back to τ decay and S-wave hyperon decay. It is not difficult to show that if we apply the CA techniques to these processes using Model (i), the results are exactly the same as those we have obtained earlier with the quark density Lagrangian. This follows from the relation

$$[Q_5^\alpha, \mathcal{L}^{(\pm),6}] = i f_{6\alpha\beta} \mathcal{L}^{(\mp),\beta},$$

where $\mathcal{L}^{(+),6}$ $[\mathcal{L}^{(-),6}]$ stands for the parity-conserving (parity-violating) part of the current-current Lagrangian of Model (i).

The situation is somewhat different with Model (ii). First observe

$$[Q_5^\alpha, \mathcal{L}_{1/2}] \propto \mathcal{L}'_{1/2} \qquad \alpha = 1,2,3$$
$$[Q_5^\alpha, \mathcal{L}_{3/2}] \propto \mathcal{L}'_{3/2} \qquad \text{"} \quad ,$$

where $\mathcal{L}'_{1/2}(\mathcal{L}'_{3/2})$ stands for some isospin transform of $\mathcal{L}_{1/2}$ $(\mathcal{L}_{3/2})$. This statement would be trivial if Q_5^α in the above commutation relation were replaced by Q^α (just isospin); if we apply an isospin rotation to $\mathcal{L}_{1/2}(\mathcal{L}_{3/2})$, we obviously get $\mathcal{L}'_{1/2}$ $(\mathcal{L}'_{3/2})$. Now in the chiral algebra there is no essential distinction between V and A as far as the isospin properties are concerned. So the above commutation relations still hold provided $\mathcal{L}_{1/2}$, $\mathcal{L}_{3/2}$ are of the current-current type.

The soft-pion techniques now tell us that if $A \rightarrow B$ is due to $\mathcal{L}_{1/2}$ ($\mathcal{L}_{3/2}$), then $A' \rightarrow B' + \pi_{soft}$ is also due to $\mathcal{L}'_{1/2}$ ($\mathcal{L}'_{3/2}$) where A (B) and A' (B') are in the same isospin multiplet.

Specializing now to $K_{\pi 3}$ (τ) decay, we can say that if $K_{\pi 2}$ decay satisfies the $\Delta T = 1/2$ rule (as is the case if we can ignore $K^+ \rightarrow \pi^+ + \pi^0$), the $\Delta T = 3/2$ amplitude in $K_{\pi 3}$ decay must vanish at three points indicated by $\times$.

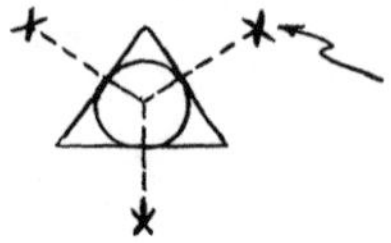

$\Delta T = 3/2$ amplitude = 0.

This means that in the linear matrix element approximation the amplitude obeys the $\Delta T = 1/2$ rule everywhere. For the $\Delta T = 1/2$ part of τ decay the predictions of Model (ii) are the same as before.

As for S wave hyperon decay we mention that Model (ii) gives neither $A(\Sigma^+_+) = 0$ nor the Lee-Sugawara relation. But we do get a set of relations such that, when supplimented by the $\Delta T = 1/2$ rule, it becomes equivalent to $A(\Sigma^+_+) = 0$ and the Lee-Sugawara relation (Sugawara '65; Suzuki '65).

Digression on Generalized Ward - Takahashi Identities

In deriving the master formula we used

$$\partial_\mu \{T(j^\alpha_{5\mu}(x) d(x))\} = T(\partial_\mu j^\alpha_{5\mu} d(0)) + \delta(x_0)[j^\alpha_{50}(x), d(0)] .$$

For discussing many-soft-π processes it is convenient to have a generalized form of this equation. Almost all low energy theorems ever proposed can be derived using the identity

$$\partial_\mu \{T(J_\mu(x) A(x') B(x'') C(x''') \ldots)\} \equiv T(\partial_\mu J_\mu(x) A(x') B(x''), \ldots) +$$
$$+ \delta(x_0 - x_0') T([J_0(x), A(x')] B(x'') C(x''') \ldots)$$
$$+ \delta(x_0 - x_0'') T(A(x')[J_0(x), B(x'')] C(x''') \ldots)$$
$$+ \ldots$$

where J_μ may stand for j^α_μ or $j^\alpha_{5\mu}$. In general, this formula can be used to relate an $n+1$ point function to an n point function. It is particularly powerful because:

i) Usually $\partial_\mu J_\mu$ either is zero (e.g. $\partial_\mu j^{(em)}_\mu = 0$) or is known (e.g. PCAC).

ii) $[J_0, A]$ is also known in many cases.

Takahashi ('57) first used this identity in QED (in the Heisenberg representation) to relate the e.m. vertex of the electron,

$$\langle 0 | T(j_\mu(x) \psi(x') \bar{\psi}(x'')) | 0 \rangle ,$$

to the electron propagator,

$$\langle 0 | T(\psi(x') \bar{\psi}(x'')) | 0 \rangle ,$$

in the $k \to 0$ limit. In this way he obtained the Ward identity (which states that the insertion of a $k = 0$ photon is equivalent to the differentiation of the electron propagator in momentum space) without recourse to perturbation theory. For this reason the formula is known as the generalized Ward-Takahashi identity.

Double Pion Emission in Strong Interaction Processes

We consider

$$A \to B + \pi_{\text{soft}} + \pi_{\text{soft}} .$$

The reduction formula gives

$$\langle B \pi^{\alpha}_{q} \pi^{\beta}_{q'} | A \rangle = i^2 \int d^4x' \int d^4x \, \frac{e^{-iq'\cdot x'}}{\sqrt{2q_0'}} \frac{e^{-iq\cdot x}}{\sqrt{2q_0}} (-\Box' + m_\pi^2)(-\Box + m_\pi^2) \times$$

$$\times \langle B | T(\varphi^{\alpha}(x) \varphi^{\beta}(x')) | A \rangle ,$$

$$\lim_{q,q' \to 0} \sqrt{4 q_0 q_0'} \langle B \pi^{\alpha}_{q} \pi^{\beta}_{q'} | A \rangle = \frac{-1}{C_\pi^2} \lim_{q,q' \to 0} \left(\frac{q'^2 + m_\pi^2}{m_\pi^2} \right) \left(\frac{q^2 + m_\pi^2}{m_\pi^2} \right) \times$$

$$\times \int d^4x' \int d^4x \, e^{-iq'\cdot x' - iq\cdot x} \langle B | T(\partial_\mu j^{\alpha}_{5\mu}(x) \, \partial_\nu j^{\beta}_{5\nu}(x')) | A \rangle .$$

It is easier to start with

$$q_\mu q_\nu' M_{\mu\nu} = q_\mu q_\nu' \int d^4x' \int d^4x \, e^{-iq'\cdot x' - iq\cdot x} \langle B | T(j^{\alpha}_{5\mu}(x) \, j^{\beta}_{5\nu}(x')) | A \rangle .$$

Write

$$-q_\mu q_\nu' \, e^{-iq'\cdot x' - iq\cdot x} = \partial_\mu \partial_\nu' \, e^{-iq'\cdot x' - iq\cdot x}$$

and integrate by parts. Treating q and q$'$ as symmetrically as possible and taking advantage of the generalized Ward-Takahashi identity, we get

$$\frac{1}{2}(\partial_\mu \partial_\nu' + \partial_\nu' \partial_\mu) T(j^{\alpha}_{5\mu}(x) \, j^{\beta}_{5\nu}(x')) =$$

$$= \frac{1}{2} \partial_\mu \left\{ T(j^{\alpha}_{5\mu}(x) \, \partial_\nu' j^{\beta}_{5\nu}(x')) + \delta(x_0 - x_0') [j^{\beta}_{50}(x'), j^{\alpha}_{5\mu}(x)] \right\}$$

$$+ \frac{1}{2} \partial_\mu' \left\{ T(\partial_\mu j^{\alpha}_{5\mu}(x) \, j^{\beta}_{5\nu}(x')) + \delta(x_0 - x_0') [j^{\alpha}_{50}(x), j^{\beta}_{5\nu}(x')] \right\}$$

$$= T(\partial_\mu j^\alpha_{5\mu}(x)\partial'_\nu j^\beta_{5\nu}(x'))$$

$$+ \frac{1}{2}\delta(x_0-x'_0)\Big([j^\alpha_{50}(x), \partial'_\nu j^\beta_{5\nu}(x')] + [j^\beta_{50}(x'), \partial_\mu j^\alpha_{5\mu}(x)]\Big)$$

$$+ \frac{1}{2}(-\partial_\mu)\Big([j^\alpha_{5\mu}(x), j^\beta_{50}(x')]\delta(x_0-x'_0)\Big)$$

$$+ \frac{1}{2}(+\partial'_\nu)\Big([j^\alpha_{50}(x), j^\beta_{5\nu}(x')]\delta(x_0-x'_0)\Big).$$

So

$$q_\mu q'_\nu M_{\mu\nu} = -\int d^4x\int d^4x'\, e^{-iq\cdot x-iq'\cdot x'}\langle B|T(\partial_\mu j^\alpha_{5\mu}(x)\partial'_\nu j^\beta_{5\nu}(x'))|A\rangle$$

$$-\frac{1}{2}\int d^4x\int d^4x'\, e^{-iq\cdot x-iq'\cdot x'}\langle B|\Big([j^\alpha_{50}(x), \partial'_\nu j^\beta_{5\nu}(x')] + [j^\beta_{50}(x'), \partial_\mu j^\alpha_{5\mu}(x)]\Big)|A\rangle\delta(x_0-x'_0)$$

$$-\frac{1}{2}\int d^4x\int d^4x'\, e^{-iq\cdot x-iq'\cdot x'}(-iq_\mu)\langle B|[j^\alpha_{5\mu}(x), j^\beta_{50}(x')]|A\rangle\delta(x_0-x'_0)$$

$$-\frac{1}{2}\int d^4x\int d^4x'\, e^{-iq\cdot x-iq'\cdot x'}(+iq'_\nu)\langle B|[j^\alpha_{50}(x), j^\beta_{5\nu}(x')]|A\rangle\delta(x_0-x'_0).$$

Use translational invariance, substitute $x \to x+x'$, integrate, and let $q, q' \to 0$ keeping only terms linear in q, q'. The second term, for example, becomes

$$\int d^4x\int d^4x'\, e^{-iq\cdot x-iq'\cdot x'}\delta(x_0-x'_0)\langle B|[j^\alpha_{50}(x), \partial'_\nu j^\beta_{5\nu}(x')]|A\rangle =$$

$$= \int d^4x\int d^4x'\, e^{-iq\cdot x-iq'\cdot x'-ip_B\cdot x'+ip_A\cdot x'}\delta(x_0-x'_0)\langle B|[j^\alpha_{50}(x-x'), \partial'_\nu j^\beta_{5\nu}(0)]|A\rangle$$

$$= (2\pi)^4 \delta^{(4)}(p_A - p_B - q - q') \int d^4x\, e^{-iq\cdot x} \delta(x_0) \langle B | [j^\alpha_{50}(x), \partial_\nu j^\beta_{5\nu}(0)] | A \rangle$$

$$= (2\pi)^4 \delta^{(4)}(p_A - p_B - q - q') \langle B | [Q^\alpha_5(0), \partial_\nu j^\beta_{5\nu}(0)] | A \rangle .$$

Since the first term is just the matrix element for double pion emission, we get the net result

$$\lim_{q, q' \to 0} \sqrt{4 q_0 q_0'} \langle \pi^\alpha_q \pi^\beta_{q'} B | A \rangle =$$

$$= \lim_{q, q' \to 0} \frac{q_\mu q'_\nu}{C_\pi^2} \int d^4x \int d^4x'\, e^{-iq\cdot x - iq'\cdot x'} \langle B | T(j^\alpha_{5\mu}(x) j^\beta_{5\nu}(0)) | A \rangle$$

$$+ (2\pi)^4 \delta^{(4)}(p_A - p_B - q - q') \frac{1}{2C_\pi^2} \Big\{ \langle B | ([Q^\alpha_5(0), \partial_\nu j^\beta_{5\nu}(0)] + [Q^\beta_5(0), \partial_\mu j^\alpha_{5\mu}(0)]) | A \rangle$$

$$+ i q_\mu \langle B | [Q^\beta_5(0), j^\alpha_{5\mu}(0)] | A \rangle + i q'_\nu \langle B | [Q^\alpha_5(0), j^\beta_{5\nu}(0)] | A \rangle \Big\} .$$

For the last two terms use, as usual,

$$[Q^\alpha_5(0), j^\beta_{5\nu}(0)] = i \varepsilon_{\alpha\beta\gamma} j^\gamma_\nu(0) .$$

Because of the appearance of $\varepsilon_{\alpha\beta\gamma}$ in the commutation relation the last two terms are antisymmetric with respect to α and β. B.E. statistics is still satisfied because the space part, $(q-q')_\mu$ is also antisymmetric. The term

$$[Q^\alpha_5(0), \partial_\nu j^\beta_{5\nu}(0)] + [Q^\beta_5(0), \partial_\mu j^\alpha_{5\mu}(0)] \equiv 2i S_{\alpha\beta}(0) \qquad (S_{\alpha\beta} = S_{\beta\alpha})$$

is symmetric in α and β, and therefore its matrix element gives no contribution to the antisymmetric amplitude. So

$$\lim_{q, q' \to 0} \sqrt{4 q_0 q_0'} \langle B (\pi^\alpha_q \pi^\beta_{q'})_{\text{antisym.}} | A \rangle = \text{"antisymmetric pole term"}$$

$$- i (2\pi)^4 \delta^{(4)}(p_A - p_B - q - q') \frac{(q-q')_\mu}{2C_\pi^2} i \varepsilon_{\alpha\beta\gamma} \langle B | j^\gamma_\mu(0) | A \rangle ,$$

which is appropriate for a p-wave (antisym. space part) pion pair.

In contrast, the symmetric amplitude is

$$\lim_{q,q'\to 0} \sqrt{4q_0 q_0'}\langle B(\pi^\alpha_q \pi^\beta_{q'})\text{sym.}|A\rangle = \text{"symmetric pole term"}$$
$$- i(2\pi)^4 \delta^{(4)}(p_A - p_B - q - q')\frac{1}{C_\pi^2}\langle B|S_{\alpha\beta}(0)|A\rangle$$

and is appropriate for an S-wave (sym. space part) pion pair.

These formulae can be used to relate A→B + ($\pi\pi$) p-wave to A→B + $\gamma^{(V)}$ (isovector photon) in the soft-pion limit (Kawarabayashi and Suzuki '66).

For the symmetric amplitude we need to know the commutator $[Q_5^\alpha, \partial_\nu j^\beta_{5\nu}]$, often called the σ term. In the PCAC language it specifies how the pion field transforms under chiral transformation. But this commutator is model dependent, and so usual current algebra tells us nothing about it:

(i) In the σ-model, $[Q_5^\alpha, \partial_\nu j^\beta_{5\nu}] \propto \delta_{\alpha\beta}\sigma$ since (π^α, σ) form a chiral quadruplet.

(ii) In the free-field quark model, $\partial_\nu j^\beta_{5\nu} \propto \bar{q}\gamma_5\lambda_\beta q$; and so (see Chapter II)

$$[Q_5^\alpha, \partial_\nu j^\beta_{5\nu}] \propto \frac{1}{3}\delta_{\alpha\beta}\bar{q}q + d_{\alpha\beta\gamma}\bar{q}\frac{\lambda_\gamma}{2}q .$$

As for the pole terms, they must have a $1/q$ (or $1/q'$) singularity in order to contribute to the soft - π - pair amplitude.

Elastic Soft-Pion Scattering

The reduction formula for scattering differs from that for double pion emission just by the sign of the four-momentum of one of the pions. In the previous formulae simply let

$$q' \longrightarrow q' \quad , \quad q \longrightarrow -q,$$

where q now stands for the initial π momentum.

$$\lim_{q,q'\to 0} \sqrt{4q_0 q_0'} \langle \pi^{\beta}_{q'} A' | \pi^{\alpha}_{q} A \rangle = \text{"pole"}$$
$$- i(2\pi)^4 \delta^{(4)}(p+q-p'-q') \frac{1}{C_\pi^2} \left\{ \langle A' | S_{\alpha\beta}(0) | A \rangle - \tfrac{1}{2}(q+q')_\mu i \varepsilon_{\alpha\beta\gamma} \langle A' | j^{\gamma}_{\mu}(0) | A \rangle \right\}$$

To determine the pole term, use the by-now-familiar argument and evaluate

according to the gradient-coupling prescription at both vertices. In the π-N case, the pole term,

$$\text{"pole"} = -i(2\pi)^4 \delta^{(4)}(p+q-p'-q') \left(\frac{G_{\pi NN}}{2m_\pi} \right)^2 \times$$

$$\times \bar{u}' \left\{ \gamma_5 \gamma \cdot q' \left(\frac{-i\gamma \cdot p + m_N}{2p \cdot q} \right) \gamma_5 \gamma \cdot q \, \tau_\beta \tau_\alpha + \gamma_5 \gamma \cdot q \left(\frac{-i\gamma \cdot p + m_N}{-2p \cdot q'} \right) \gamma_5 \gamma \cdot q' \, \tau_\alpha \tau_\beta \right\} u$$

gives no contribution to the s wave in the q, $q' \longrightarrow 0$ limit.

The symmetric $\langle A | S_{\alpha\beta} | A \rangle$ term is expected to be small for the following reasons:

(1) If we use a formalism in which $\partial_\mu j^{\alpha}_{5\mu} = 0$ and $m_\pi^2 = 0$, then this term is automatically absent. So it is of order $\mathcal{O}(m_\pi^2/m_A^2)$.

(ii) In the σ model it is again proportional to m_π^2 since

$$[Q_5^\alpha, \partial_\mu j_{5\mu}^\beta] = c_\pi m_\pi^2 [Q_5^\alpha, \pi^\beta] = \left(\frac{g_A}{g_V}\right) c_\pi m_\pi^2 \delta_{\alpha\beta}\sigma.$$

(iii) Adler's consistency condition plus the smoothness (of the q^2 dependence) assumption says that the symmetric part of the π-N scattering amplitude at threshold is of order $\mathcal{O}(m_\pi^2/m_A^2)$. This would be hard to understand if $\langle A|S_{\alpha\beta}|A\rangle$ were large. (Note that the off-mass-shell amplitude we are considering depends on both q^2 and q'^2. Previously Adler's condition refers to the off-mass-shell amplitude where only one of them, say q^2, is varied. PCAC presumably requires that the amplitude is smooth in both q^2 and q'^2.)

Thus only the vector term arising from the Gell-Mann commutator appears sizable. Assume that the momentum dependence of the on-mass-shell amplitude at threshold is still correctly given by this term alone. This gives the π-A S-wave scattering length,

$$a_T = -\left(\frac{m_\pi m_A}{m_\pi + m_A}\right)\frac{1}{4\pi C_\pi^2}\vec{T}_\pi\cdot\vec{T}_A + \mathcal{O}(m_\pi^2/m_A^2),$$

$$\vec{T}_\pi\cdot\vec{T}_A = \frac{1}{2}\left[T(T+1) - T_A(T_A+1) - 2\right],$$

where we have used $i\,\varepsilon_{\alpha\beta\gamma} = -(T_\pi^\gamma)_{\alpha\beta}$. This form was first written down by Weinberg ('66), although special cases were noted earlier by Tomozawa ('66) and others. In the particular case of π-N scattering there is very good agreement with experiment, as the following table shows.

	Experiment	CA + PCAC prediction	
$a_1 + 2a_3$	$(0.00 \pm 0.01)/m_\pi$	0	(0)
$a_1 - a_3$	$(0.27 \pm 0.01)/m_\pi$	$0.24/m_\pi$	$(0.30/m_\pi)$
	using C_π from	π decay	(G - T rel.)

Note that $a_1 + 2a_3$ tests the smallness of the σ term, while $a_1 - a_3$ directly tests the Gell-Mann commutation relations.

Adler-Weisberger Relation

The antisymmetric combination $a_1 - a_3$ can be written as an integral over the toal $\pi^\pm$ p cross section difference using the 1955 dispersion-relation sum rule of Goldberger, Miyazawa, and Oehme,

$$D^{(-)}(m_\pi) = \frac{G^2_{\pi NN}\, m_\pi}{4\pi m_N^2} + \frac{m_\pi}{2\pi^2}\int_{m_\pi}^{\infty} d\omega\, \frac{\sigma^{(-)}(\omega)}{|\vec{q}|},$$

where $D^{(-)}(\omega)$ = real part of the antisymmetric part of the forward π-N scattering amplitude in the lab. system,

ω = lab. π energy,

$|\vec{q}| = (\omega^2 - m_\pi^2)^{1/2}$ = lab. π momentum,

$$\sigma^{(-)}(\omega) = \sigma^{(\pi^- p)}(\omega) - \sigma^{(\pi^+ p)}(\omega).$$

Multiply the current algebra prediction,

$$D^{(-)}(m_\pi) = \frac{2}{3}\left(1 + \frac{m_\pi}{m_N}\right)(a_1 - a_3) = \frac{m_\pi}{4\pi C_\pi^2},$$

by (assuming the G-T rel. is exact)

$$\frac{4\pi C_\pi^2}{m_\pi}\left(\frac{g_V}{g_A}\right)^2 = \frac{4\pi}{m_\pi}\left(\frac{m_N^2}{G^2_{\pi NN}}\right)$$

and rewrite the dispersion sum rule for $D^{(-)}(m_\pi)$ using

$$W^2 = (\text{c.m. energy})^2 = (m_N + \omega)^2 - |\vec{q}|^2$$
$$= m_N^2 + m_\pi^2 + 2m_N\omega ,$$

$$|\vec{q}| = \left[\frac{(W^2 - m_N^2 - m_\pi^2)^2}{4m_\pi^2} - m_\pi^2\right]^{1/2} \approx \frac{1}{2m_N}(W^2 - m_N^2),$$

$$W\,dW = m_N\,d\omega .$$

We then get the result

$$1 - \left(\frac{g_V}{g_A}\right)^2 = -\left(\frac{4m_N^2}{G_{\pi NN}^2}\right)\frac{1}{\pi}\int_{m_N+m_\pi}^{\infty} dW\,\frac{W\,\sigma^{(-)}(W)}{W^2 - m_N^2} .$$

This is the celebrated Adler-Weisberger relation, which was first obtained using the Fubini-Furlan technique discussed in Chapter II. (Adler '65; Weisberger '65). Numerically from the measured $\pi^{\pm}p$ total cross sections we get

$$1 - (g_V/g_A)^2 = 0.246$$

or

$$|g_A/g_V| = 1.15.$$

This is in excellent agreement with

$$-g_A/g_V = 1.18 \pm 0.02$$

determined directly from nucleon β decay.

An Adler-Weisberger type relation can also be written down for K particle processes by applying PCAC to kaons. The cross section integral now involves $\sigma(K^{\pm}p)$ or $\sigma(K^{\pm}n)$. In this way it is possible to estimate the D/F ratio for the axial vector part of baryon β decays. We get (Amati, Bouchiat,

Nuyts '65; Levinson and Muzinich '65; Weisberger '66)

$$\alpha = \frac{D}{D + F} \approx 0.6 \text{ —— } 0.8$$

in fair agreement with other estimates ($\alpha = 0.64 \pm 0.03$ from the leptonic decays of hyperons).

<u>Connection with Vector Meson Dominance</u> (Sakurai '67).

In the scattering length formula which we obtained from CA

$$a_T = - \vec{T}_\pi \cdot \vec{T}_A \times (\text{universal const.})$$

the sign is such that there is repulsion for parallel isospins and attraction for antiparallel isospins.

But this $- \vec{T}_\pi \cdot \vec{T}_A$ dependence is just what would be expected on the basis of ρ dominance,

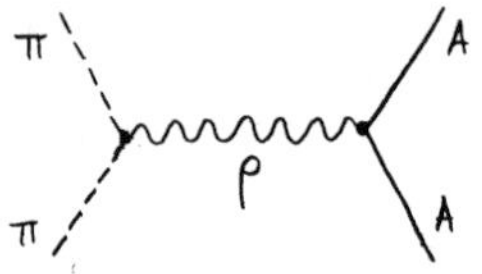

where ρ is assumed to be coupled universally to the isospin.

This similarity is no accident. To see the connection go back to the current-field identity,

$$j_\mu^\alpha = \frac{m_\rho^2}{f_\rho} \rho_\mu^\alpha .$$

In deriving the scattering length formula we used PCAC but not CFI. From CA + PCAC <u>and CFI</u> we get

$$\mathcal{M}(\pi \text{ soft} + A \to \pi \text{ soft} + A) = \frac{-1}{2C_\pi^2} (g + g')_\mu \, i \varepsilon_{\alpha\beta\gamma} \frac{m_\rho^2}{f_\rho} \frac{\langle A | J_\mu^{(\rho),\gamma} | A \rangle}{m_\rho^2 - t} .$$

This says that soft-pion scattering is due to ρ propagation in the t-channel. From this point of view, CA justifies the 1960 proposal that the whole of the S-wave π-N scattering at threshold is due to ρ exchange. More quantitatively, compare this with the ρ exchange amplitude

$$\mathcal{M} = - f_{\rho\pi\pi}(q+q')_\mu \, i\, \varepsilon_{\alpha\beta\gamma} \left. \frac{\langle A | J^{(\rho),\gamma} | A \rangle}{m_\rho^2 - t} \right|_{t=0}$$

to get the coupling constant relation,

$$f_\rho f_{\rho\pi\pi} = \frac{m_\rho^2}{2C_\pi^2} \, .$$

If we further assume $f_\rho = f_{\rho\pi\pi}$, then

$$\frac{f_\rho^2}{4\pi} = \frac{m_\rho^2}{8\pi C_\pi^2} \quad , \text{ where } \begin{cases} f_\rho^2/4\pi = 2.5 \pm 0.1 \\ \dfrac{m_\rho^2}{8\pi C_\pi^2} = 2.66 \end{cases} .$$

This (KSRF) relation was first obtained by Kawarabayashi and Suzuki ('66) and Riazuddin and Fayyazuddin ('66) using somewhat different arguments.

The connection between CA and vector meson dominance is also made apparent in other areas. For example, the successful predictions of CA regarding S-wave hyperon decay are completely identical to the predictions of the K^* dominance model of Lee and Swift ('64), based on the following diagram.

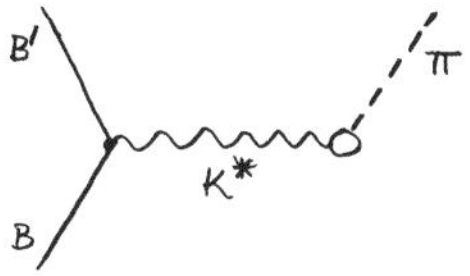

VI. ASYMPTOTIC SYMMETRIES AND SPECTRAL-FUNCTION SUM RULES

Asymptotic Chiral Symmetry and the Weinberg Mass Relation

If chiral $SU(2) \otimes SU(2)$ were an exact symmetry, we would expect an axial vector meson ($T=1$, $J^{PG}=1^{+-}$) degenerate with $\rho(774)$. The only possible candidate is the $A_1(1080)$ meson ($\rho\pi$ resonance consistent with 1^+ assignment). But $m_{A_1}^2 \approx 2m_\rho^2$, so clearly this is a "broken" symmetry. However, we may conjecture that broken symmetries such as chiral $SU(2) \otimes SU(2)$ become exact at high momentum (or equivalently at short distances), since here large mass differences might not matter so much.

Suppose we consider an external field W_μ^α (not quantized) coupled to j_μ^α.

$$L = W_\mu^\alpha \, j_\mu^\alpha \qquad\qquad \alpha = 1, 2, 3$$

One manifestation of charge independence (SU(2)) is that the vacuum to vacuum amplitude in the presence of

$$\left.\begin{aligned} W_\mu^1 &= f_\mu(x) \\ W_\mu^{2,3} &= 0 \end{aligned}\right\} \text{ is the same as with } \left\{\begin{aligned} W_\mu^{1,2} &= 0 \\ W_\mu^3 &= f_\mu(x) . \end{aligned}\right.$$

In other words, the physics is independent of the "coordinate system" in isospin space.

Let us now consider

$$L = W^\alpha_\mu J^\alpha_\mu + W^\alpha_{5\mu} J^\alpha_{5\mu} .$$

If $SU(2) \otimes SU(2)$ were exact, we would expect that the vacuum to vacuum amplitude in the presence of $W^\alpha_\mu = f^\alpha_\mu(x)$ is the same as with $W^\alpha_{5\mu} = f^\alpha_\mu(x)$. To see whether this is so, we work within the framework of the pole approximation in which only ρ for the vector current and A_1 and π for the axial current are kept.

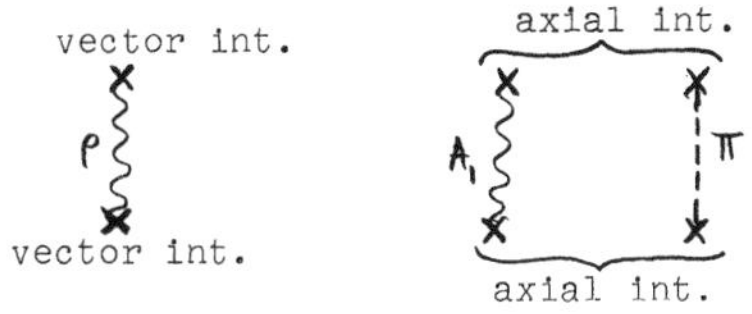

Vector case:

$$\tilde{W}^\alpha_\mu(q)\left[\left(\frac{m_\rho^2}{f_\rho}\right)^2 \frac{(\delta_{\mu\nu} + \frac{q_\mu q_\nu}{m_\rho^2})}{q^2 + m_\rho^2 - i\epsilon}\right]\tilde{W}^\alpha_\nu(q) ,$$

Axial case:

$$\tilde{W}^\alpha_{5\mu}(q)\left[\left(\frac{m_A^2}{f_A}\right)^2 \frac{(\delta_{\mu\nu} + \frac{q_\mu q_\nu}{m_A^2})}{q^2 + m_A^2 - i\epsilon} + \frac{c_\pi^2\, q_\mu q_\nu}{q^2 + m_\pi^2 - i\epsilon}\right]\tilde{W}^\alpha_{5\mu}(q) ,$$

where $\tilde{W}^\alpha_{\mu,5\mu}$ in the Fourier transform of $W^\alpha_{\mu,5\mu}$, and m_A^2/f_A is defined in the same way as m_ρ^2/f_ρ, i.e.,

$$\langle 0 | j^\alpha_{5\mu}(0) | A_1 \rangle = \frac{m_A^2}{f_A} \frac{1}{\sqrt{2\omega}} \epsilon_\mu .$$

For exact symmetry the expressions in the brackets must be equal for all q. But this is possible only for $m_A = m_\rho$, $c_\pi = 0$, $f_A = f_\rho$. Now try a weaker form of symmetry: At high q the two expressions become identical. In other words, when the hadronic world is probed by external fields which have only high-q Fourier components, the vector and axial currents are indistinguishable. So set the coefficients of the $q_\mu q_\nu/q^2$ equal:

$$\left(\frac{m_\rho}{f_\rho}\right)^2 = \left(\frac{m_A}{f_A}\right)^2 + c_\pi^2 .$$

We may further assume that chiral symmetry is so good that even the coefficients of the $\delta_{\mu\nu}/q^2$ terms become also equal for high q. We then get

$$\left(\frac{m_\rho^2}{f_\rho}\right)^2 = \left(\frac{m_A^2}{f_A}\right)^2 .$$

Now combine the two relations as follows

$$\left(\frac{m_\rho}{f_\rho}\right)^2 = \left(\frac{m_A^2}{f_A}\right)^2 \frac{1}{m_A^2} + c_\pi^2 = \frac{m_\rho^2}{f_\rho^2} \frac{m_\rho^2}{m_A^2} + c_\pi^2$$

and use the KSRF relation, $c_\pi^2 = m_\rho^2/2f_\rho^2$. With this we get the mass relation,

$$m_A^2 = 2m_\rho^2 ,$$

first obtained by Weinberg ('67) (using a somewhat different argument). The experimental masses are in extraordinary agreement with this result.

$$m_A = (1.41 \pm 0.01) m_\rho$$

Clearly it is important to verify that the 1080 MeV bump in the $\rho\pi$ spectrum is indeed due to a T=1, $J^{PG} = 1^{+-}$ meson.

Spectral Function Sum Rules

We now treat the above problem without using pole dominance. Start with the Fourier transforms of the time-ordered products,

$$\Delta^{(V)}_{\mu\nu}(q) \equiv i \int d^4x \, e^{-iq\cdot x} \langle 0 | T(j^{\alpha}_{\mu}(x) \, j^{\alpha}_{\nu}(0)) | 0 \rangle$$

(α = 1,2,3, not summed)

$$\Delta^{(A)}_{\mu\nu}(q) \equiv i \int d^4x \, e^{-iq\cdot x} \langle 0 | T(j^{\alpha}_{5\mu}(x) \, j^{\alpha}_{5\nu}(0)) | 0 \rangle .$$

Write the standard spectral representation for $\Delta^{(V)}_{\mu\nu}(q)$, where the currents are conserved and only spin-one objects contribute.

$$\Delta^{(V)}_{\mu\nu}(q) = \int_0^\infty dm^2 \rho^{(V)}(m^2) \left(\frac{\delta_{\mu\nu} + \frac{q_\mu q_\nu}{m^2}}{q^2 + m^2 - i\varepsilon} \right),$$

where

$$\left(\delta_{\mu\nu} - \frac{p_\mu p_\nu}{p^2}\right) \rho^{(V)}(-p^2) = (2\pi)^3 \sum_M \delta^{(4)}(p - p_M) \langle 0 | j^3_\mu(0) | M \rangle \langle M | j^3_\nu(0) | 0 \rangle .$$

But this is not quite right, the reason being that the time-ordered product is not covariant, whereas the spectral representation is. To show this, define

$$\Theta_{\mu\nu} \equiv T(j_\mu^\alpha(x)\, j_\nu^\alpha(0)).$$

Then since the vector currents are conserved,

$$\partial_\mu \Theta_{\mu\nu} = \delta(x_0)[j_0^\alpha(x), j_\nu^\alpha(0)] = \begin{cases} 0 & \text{for } \nu = 4 \\ -i C_{\alpha\alpha} \partial_k \delta^{(4)}(x) & \text{for } \nu = k \quad \text{(Schwinger term).} \end{cases}$$

Thus $\partial_\mu \Theta_{\mu\nu}$ is not a 4-vector, and therefore $\Theta_{\mu\nu}$ does not transform as a second rank tensor. To remedy this, define

$$\Theta'_{\mu\nu} \equiv T(j_\mu^\alpha(x)\, j_\nu^\alpha(0)) - i C_{\alpha\alpha} \delta_{\mu 4} \delta_{\nu 4} \delta^{(4)}(x),$$

which is covariant because

$$\partial_\mu \Theta'_{\mu\nu} = \begin{cases} 0 - i C_{\alpha\alpha} \partial_4 \delta^{(4)}(x) & \text{for } \nu = 4 \\ -i C_{\alpha\alpha} \partial_k \delta^{(4)}(x) + 0 & \text{for } \nu = k \end{cases}$$

does transform as a 4-vector. This suggests that it is $\Theta'_{\mu\nu}$ rather than $\Theta_{\mu\nu}$ that has a covariant spectral representation. Meanwhile, from the Goto-Imamura argument given in Chapter 2

$$C_{\alpha\alpha} = + \int_0^\infty dm^2 \, \frac{\rho^{(v)}(m^2)}{m^2}.$$

(If the Schwinter term is not a c-number, the LHS must be replaced by the vacuum expectation value of $c_{\alpha\alpha}$.) Hence

$$\Delta^{(V)}_{\mu\nu}(q) = \int_0^\infty dm^2 \rho^{(V)}(m^2) \left(\frac{\delta_{\mu\nu} + \frac{q_\mu q_\nu}{m^2}}{q^2+m^2-i\varepsilon} \right) - \delta_{\mu 4}\delta_{\nu 4} \int_0^\infty dm^2 \frac{\rho^{(V)}(m^2)}{m^2} .$$

Likewise,

$$\Delta^{(A)}_{\mu\nu}(q) = \int_0^\infty dm^2 \rho^{(A)}(m^2) \left(\frac{\delta_{\mu\nu} + \frac{q_\mu q_\nu}{m^2}}{q^2+m^2-i\varepsilon} \right) + \frac{C_\pi^2 q_\mu q_\nu}{q^2+m_\pi^2-i\varepsilon}$$

$$- \delta_{\mu 4}\delta_{\nu 4} \left[\int_0^\infty dm^2 \frac{\rho^{(A)}(m^2)}{m^2} + C_\pi^2 \right],$$

where we have assumed for simplicity that the spin-zero contribution is due to the single π state only.

As an exercise, convince yourself that without the extra term the time-ordered product, $\langle 0|T(j_4^\alpha(x)\, j_4^\alpha(0))|0\rangle$, could not be continuous at $x_0 = 0$. (Hint: Take advantage of the fact that $\partial_4 \Delta_F(x)$ is discontinuous at $x_0 = 0$.)

Weinberg's spectral-function sum rules can now be derived by requiring the asymptotic symmetry condition (Das, Mathur and Okubo, '67):

$$\lim_{q\to\infty} \left(\Delta^{(V)}_{\mu\nu}(q) - \Delta^{(A)}_{\mu\nu}(q^2) \right) = 0 .$$

From the coefficient of the $q_\mu q_\nu / q^2$ term we get Weinberg's 1st sum rule,

$$\int_0^\infty dm^2 \frac{\rho^{(V)}(m^2)}{m^2} = \int_0^\infty dm^2 \frac{\rho^{(A)}(m^2)}{m^2} + C_\pi^2 .$$

Asymptotic symmetry may also set in at lower values of q^2. Then from the coefficient of the $\delta_{\mu\nu}/q^2$ term we get Weinberg's 2nd sum rule,

$$\int_0^\infty dm^2 \rho^{(V)}(m^2) = \int_0^\infty dm^2 \rho^{(A)}(m^2) .$$

If we now saturate the spectral functions by ρ and A_1, viz.

$$\rho^{(V)}(m^2) = \left(\frac{m_\rho^2}{f_\rho}\right)^2 \delta(m^2 - m_\rho^2)$$

$$\rho^{(A)}(m^2) = \left(\frac{m_A^2}{f_A}\right)^2 \delta(m^2 - m_A^2),$$

we have the same results as before.

Derivation from Current Algebra

With certain restrictions on the nature of the Schwinger terms the first sum rule can also be derived from the unintegrated current- commutation relations of chiral SU(2) $\otimes$ SU(2). We begin by considering $i\varepsilon_{\alpha\beta\gamma}[j_0^\gamma(x), j_k^\delta(0)]$. Taking advantage of the current-commutation relations and the Jacobi identity, we get

$$i\varepsilon_{\alpha\beta\gamma}[j_0^\gamma(x), j_k^\delta(0)] = [[Q_5^\alpha, j_{50}^\beta(x)], j_k^\delta(0)] =$$

$$= -[[j_{50}^\beta(x), j_k^\delta(0)], Q_5^\alpha] - [[j_k^\delta(0), Q_5^\alpha], j_{50}^\beta(x)] =$$

$$= [(i\varepsilon_{\delta\beta\epsilon} j_{5k}^\epsilon(x)\delta^{(3)}(\vec{x}) + S.T.), Q_5^\alpha] + i\varepsilon_{\alpha\delta\epsilon}[j_{5k}^\epsilon(0), j_{50}^\beta(x)],$$

where S.T. stands for the Schwinger term associated with the commutator $[j_{50}^{\beta}(x), j_{k}^{\delta}(0)]$. Let us now take the vacuum expectation values to obtain

$$i\varepsilon_{\alpha\beta\delta}\langle 0|[j_0^{\delta}(x), j_k^{\delta}(0)]|0\rangle =$$

$$= i\,\varepsilon_{\alpha\delta\beta}\langle 0|[j_{5k}^{\beta}(0), j_{50}^{\beta}(x)]|0\rangle + \langle 0|[(S.T.), Q_5^{\alpha}]|0\rangle,$$

where the δ indices (β indices) are not summed. The last term vanishes if

i) The Schwinger term is a q-number that transforms like isoscalar (in particular no T=1 component).

ii) The Schwinger term is zero.

(Note that the Schwinger terms for $[j_0, j_k]$ and $[j_{50}, j_{5k}]$ cannot vanish but the Schwinger term for $[j_{50}, j_k]$ can be zero; in fact, it is zero in the gauge field algebra. Note also that the case where the Schwinger term is a c-number need not be considered here because of parity conservation.) We now set $\beta \neq \delta$ (so that $\varepsilon_{\alpha\beta\delta} \neq 0$) and obtain

$$\langle 0|[j_0^{\delta}(x), j_k^{\delta}(0)]|0\rangle = \langle 0|[j_{50}^{\beta}(x), j_{5k}^{\beta}(0)]|0\rangle.$$

(Having obtained this identity, we can now relax $\beta \neq \delta$ since isospin is supposed to be conserved.) The Goto-Imamura argument, as applied to the above relation, immediately implies the first sum rule of Weinberg. From this point of view, the first sum rule is equivalent to the statement that the Schwinger terms (or the vacuum expectation values of the Schwinger terms) of the VV and the AA commutators are equal.

Electromagnetic Mass Difference of Pions

In the soft-pion limit, $m_{\pi^+} - m_{\pi^0}$ is calculable in terms of $\rho^{(V)}$ and $\rho^{(A)}$. If we use the Weinberg sum rules, the result is convergent and agrees well with observation, as shown by Das, Guralnik, Mathur, Low and Young ('67).

The e.m. mass difference is related to Compton scattering of a virtual photon on a soft pion

The first diagram is due to the $\varphi^2 A_\mu^2$ interaction. But if we use a gauge in which the photon propagator is

$$\frac{1}{q^2 - i\varepsilon}\left(\delta_{\mu\nu} - 4\frac{q_\mu q_\nu}{q^2}\right),$$

then since

$$\delta_{\mu\nu}\left(\delta_{\mu\nu} - \frac{4 q_\mu q_\nu}{q^2}\right) = 4 - 4\frac{q^2}{q^2} = 0,$$

the A_μ^2 term does not contribute. In this gauge the mass difference can be written as

$$m_{\pi^+}^2 - m_{\pi^0}^2 = -\frac{e^2}{4\pi}\, 2m_\pi \,\mathrm{Re}\int \frac{d^4q}{q^2 - i\varepsilon}\left(\delta_{\mu\nu} - \frac{4q_\mu q_\nu}{q^2}\right)\times$$

$$\times \int d^4x\, e^{iq\cdot x}\left\{\langle \pi^+|T(j_\mu^{em}(x)\, j_\nu^{em}(0))|\pi^+\rangle - \langle \pi^0|T(j_\mu^{em}(x)\, j_\nu^{em}(0))|\pi^0\rangle\right\}.$$

Actually this expression contains noncovariant pieces but they cancel anyway by virtue of Weinberg's 1st sum rule. (More properly, we should start with just the covariant parts of the T products.) Starting with

$$\partial'_\lambda \partial''_\sigma \langle 0|T(j_{5\lambda}^{1-i2,3}(x')\, j_{5\sigma}^{1+i2,3}(x'')\, j_\mu^{em}(x)\, j_\nu^{em}(0))|0\rangle$$

relate

$$\langle \pi^{+,0}|T(j_\mu^{em}(x)\, j_\nu^{em}(0))|\pi^{+,0}\rangle$$

to

$$\langle 0|T(j_\mu^\alpha(x)\, j_\nu^\alpha(0))|0\rangle \text{ and } \langle 0|T(j_{5\mu}^\alpha(x)\, j_{5\nu}^\alpha(0))|0\rangle$$

using PCAC and CA with the soft-pion technique. As for the other terms, they either vanish in the soft-pion limit or cancel (such as the σ terms) for the π^+-π^0 mass difference. As a result, it is possible to show

$$m_{\pi^+}^2 - m_{\pi^0}^2 = -i\frac{e^2}{(2\pi)^4}\frac{1}{C_\pi^2}\int \frac{d^4q}{q^2 - i\varepsilon}\left(\delta_{\mu\nu} - \frac{4q_\mu q_\nu}{q^2}\right)\left[\Delta_{\mu\nu}^{(V)}(q) - \Delta_{\mu\nu}^{(A)}(q)\right].$$

With our particular choice of gauge, only the $q_\mu q_\nu$ part of $\Delta_{\mu\nu}^{(V,A)}$ contributes.

$$\left(\delta_{\mu\nu} - \frac{4q_\mu q_\nu}{q^2}\right)\begin{Bmatrix}\delta_{\mu\nu}\\ q_\mu q_\nu\end{Bmatrix} = \begin{Bmatrix}0\\ -3q^2\end{Bmatrix}$$

Furthermore, rewrite the spectral representation for $\Delta_{\mu\nu}^{(V,A)}$ using

$$\frac{1}{q^2+m^2} = \frac{1}{q^2} - \frac{m^2}{q^2(q^2+m^2)} .$$

Ignoring the m_π^2 term (set $m_\pi=0$), the most divergent (quadratic) parts of the $\pi^+-\pi^0$ mass difference are now seen to cancel by virtue of Weinberg's 1st sum rule. We are left with

$$m_{\pi^+}^2 - m_{\pi^0}^2 = -3i \frac{e^2}{(2\pi)^4} \frac{1}{c_\pi^2} \int \frac{d^4q}{q^2} \int_0^\infty dm^2 \, \frac{\rho^{(V)}(m^2) - \rho^{(A)}(m^2)}{q^2+m^2} .$$

This still looks logarithmically divergent; but because of Weinberg's 2nd sum rule, it is completely convergent. Now make the pole approximation which we used to obtain Weinberg's mass relation. Integrate à la Feynman, and we get the result

$$m_{\pi^+}^2 - m_{\pi^0}^2 = \frac{3 \ln 2}{2\pi} \frac{e^2}{4\pi} m_\rho^2$$

$$m_{\pi^+} - m_{\pi^0} \approx 5.0 \text{ Mev} ,$$

in reasonable agreement with the experimental value of 4.6 MeV. It is amusing that c_π^2 drops out in the final expression.

Comparison of Other Currents; $\omega\varphi$ Mixing

If there is an axial K^* meson, we can estimate its mass in terms of the $K^*(890)$ mass and the $K_{\mu 2}$ and K_{e3} decay constants. By comparing

$$\langle 0|T(j^4_\mu j^4_\nu)|0\rangle \text{ and } \langle 0|T(j^4_{5\mu} j^4_{5\nu})|0\rangle$$

Das, Mathur and Okubo ('67) got $K^*_A = 1311$ MeV. This corresponds well with a $K\pi\pi$ resonance (most likely 1^+) at 1313 ± 8 MeV.

Perhaps this approach also makes sense for the broken eight-fold way; so compare

$$\langle 0|T(j^3_\mu j^3_\nu)|0\rangle \text{ and } \langle 0|T(j^8_\mu j^8_\nu)|0\rangle .$$

Saturating the sum rules in this case with ρ, ω and φ, we get

1st sum rule: $$\frac{m_\rho^2}{f_\rho^2} = \frac{3}{4}\left(\frac{\cos^2\theta_Y}{f_Y^2} m_\varphi^2 + \frac{\sin^2\theta_Y}{f_Y^2} m_\omega^2\right),$$

2nd sum rule: $$\frac{m_\rho^4}{f_\rho^2} = \frac{3}{4}\left(\frac{\cos^2\theta_Y}{f_Y^2} m_\varphi^4 + \frac{\sin^2\theta_Y}{f_Y^2} m_\omega^4\right).$$

Eliminating the coupling constants, we get

$$\tan^2\theta_Y = \left(\frac{m_\varphi^2}{m_\omega^2}\right)\frac{(m_\rho^2 - m_\varphi^2)}{(m_\omega^2 - m_\rho^2)} .$$

This relation is catastrophic! The LHS must be positive while the RHS is negative (since $m_\varphi > m_\omega > m_\rho$). Perhaps we can be a little generous and admit the possibility

$m_\omega = m_\rho$ since experimentally ω and ρ are nearly degenerate. This is still bad because it implies $\Theta_Y = \pi/2$ and no $\omega - \varphi$ mixing, hence $\varphi \not\rightarrow K^+K^-$ and $\varphi \not\rightarrow e^+e^-$. So the 2nd sum rule is no good, or else vector meson dominance is "out," at least for the 2nd sum rule.

As a direct test of the 1st sum rule + vector meson dominance, we may mention the following relation among the leptonic decays of vector mesons, which is independent of the details of $\omega - \varphi$ mixing (Das, Mathur and Okubo '67):

$$\frac{1}{3} m_\rho \Gamma_{\rho \to \ell^+\ell^-} = m_\omega \Gamma_{\omega \to \ell^+\ell^-} + m_\varphi \Gamma_{\varphi \to \ell^+\ell^-}.$$

Incidentally, if you believe in the 1st sum rule but not in vector meson dominance, you can still write down a sum rule involving colliding beam cross sections

$$\frac{1}{3} \int ds\, s\, \sigma(e^+ + e^- \longrightarrow T = 1 \text{ hadronic system})$$

$$= \int ds\, s\, \sigma(e^+ + e^- \longrightarrow T = 0 \text{ hadronic system})$$

It is also possible to show, by requiring that the Fourier transform of $\langle 0 | T(j_\mu^{(B)}(x)\, j_\nu^{(Y)}(0)) | 0 \rangle$ go to zero as $q \to \infty$, that the current-mixing (Coleman-Schnitzer) model rather than the mass-mixing model should be used to characterize the ω-φ complex (Oakes and Sakurai '67). Combining the spectral function sum rules with the ω-φ mixing scheme of Coleman and Schnitzer, we can relate the coupling constants for $\gamma \leftrightarrow \rho, \omega, \varphi$ in the broken eightfold way. The dimensionless constants which would satisfy the 9:1:2 ratio in the SU(6) limit now obey

$$\frac{1}{f_\rho^2} : \frac{\sin^2\theta_Y}{4f_Y^2} : \frac{\cos^2\theta_Y}{4f_Y^2} = 9.00 : 0.65 : 1.33 .$$

Finally compare $\langle 0|T(j_\mu^3 j_\nu^3)|0\rangle$ and $\langle 0|T(j_\mu^4 j_\nu^4)|0\rangle$. The 1st and 2nd sum rules require that either $m_{K*} = m_\rho$ (which is bad) or we have a scalar (0^+) κ meson associated with the nonconserved strangeness-changing current. My personal view is that only the 1st sum rule is satisfied in the broken eightfold way whereas both sum rules are good for chiral SU(2) $\otimes$ SU(2).

APPENDIX, NOTATION AND NORMALIZATION

The γ matrices and the metric

See my previous two books: Invariance Principles and Elementary Particles (Princeton University Press, 1964), Advanced Quantum Mechanics (Addison-Wesley Publishing Co., 1967).

The λ matrices

$$\left.\begin{aligned} [\lambda_\alpha, \lambda_\beta] &= 2if_{\alpha\beta\gamma}\lambda_\gamma , \\ \{\lambda_\alpha, \lambda_\beta\} &= \tfrac{4}{3}\delta_{\alpha\beta} + 2d_{\alpha\beta\gamma}\lambda_\gamma \\ \mathrm{Tr}(\lambda_\alpha\lambda_\beta) &= 2\delta_{\alpha\beta} \end{aligned}\right\} \quad \alpha,\beta,\gamma=1,\ldots 8 .$$

The $f_{\alpha\beta\gamma}$ are real and totally antisymmetric.
The $d_{\alpha\beta\gamma}$ are real and totally symmetric.
Nonzero elements of $f_{\alpha\beta\gamma}$

$\alpha\beta\gamma$	$f_{\alpha\beta\gamma}$
123	1
147	1/2
156	-1/2
246	1/2
257	1/2
345	1/2
367	-1/2
458	$\sqrt{3}/2$
678	$\sqrt{3}/2$

Nonzero elements of $d_{\alpha\beta\gamma}$

$\alpha\beta\gamma$	$d_{\alpha\beta\gamma}$
118	$1/\sqrt{3}$
146	$1/2$
157	$1/2$
228	$1/\sqrt{3}$
247	$-1/2$
256	$1/2$
338	$1/\sqrt{3}$
344	$1/2$
355	$1/2$
366	$-1/2$
377	$-1/2$
448	$-1/(2\sqrt{3})$
558	$-1/(2\sqrt{3})$
668	$-1/(2\sqrt{3})$
778	$-1/(2\sqrt{3})$
888	$-1/\sqrt{3}$

Sometimes it is convenient to introduce

$$\lambda_o = \sqrt{2/3} \quad \text{(times the identity matrix)}$$

so that

$$\{\lambda_\alpha, \lambda_\beta\} = 2d_{\alpha\beta\gamma}\lambda_\gamma \;, \quad \alpha,\beta,\gamma = 0,1, \ldots 8$$

with

$$d_{\alpha\beta o} = \sqrt{2/3}\;\delta_{\alpha\beta} \;, \text{ etc.}$$

Current commutation relations

Once integrated relations

$$[Q^{\alpha}(x_0),\ j_{\mu}^{\beta}(x)] = if_{\alpha\beta\gamma} j_{\mu}^{\gamma}(x)$$

$$[Q^{\alpha}(x_0),\ j_{5\mu}^{\beta}(x)] = if_{\alpha\beta\gamma} j_{5\mu}^{\gamma}(x)$$

$$[Q_5^{\alpha}(x_0),\ j_{\mu}^{\beta}(x)] = if_{\alpha\beta\gamma} j_{5\mu}^{\gamma}(x)$$

$$[Q_5^{\alpha}(x_0),\ j_{5\mu}^{\beta}(x)] = if_{\alpha\beta\gamma} j_{\mu}^{\gamma}(x)$$

where the charges (Q^{α}, Q_5^{α}) and currents (j_{μ}^{α}, $j_{5\mu}^{\alpha}$) are normalized so that in the quark model we have

$$j_{\mu}^{\alpha} = i\bar{q}\gamma_{\mu}\frac{\lambda_{\alpha}}{2} q, \qquad j_{5\mu}^{\alpha} = i\bar{q}\gamma_{\mu}\gamma_5\frac{\lambda_{\alpha}}{2} q,$$

$$Q^{\alpha} = \int j_o^{\alpha}\, d^3x = -i\int j_4^{\alpha}\, d^3x ,$$

$$Q_5^{\alpha} = \int j_{5o}^{\alpha} d^3x = -i\int j_{54}^{\alpha}\, d^3x .$$

The electromagnetic interaction picks

$$j_{\mu}^{(em)} = j_{\mu}^{3} + \frac{1}{\sqrt{3}}\, j_{\mu}^{8} .$$

The semileptonic weak interaction picks

$$j_{\mu}^{(weak)} = \cos\theta \left[j_{\mu}^{1\pm i2} + j_{5\mu}^{1\pm i2} \right] + \sin\theta \left[j_{\mu}^{4\pm i5} + j_{5\mu}^{4\pm i5} \right]$$

with

$$\sin\theta = 0.21 — 0.28.$$

Meson fields

"Vector" notation

$$\eta = \varphi^8 \qquad\qquad \pi^\pm = \frac{1}{\sqrt{2}}(\varphi^1 \mp i\varphi^2)$$

$$\pi^0 = \varphi^3 \qquad\qquad K^\pm = \frac{1}{\sqrt{2}}(\varphi^4 \mp i\varphi^5)$$

$$K^0 = \frac{1}{\sqrt{2}}(\varphi^6 - i\varphi^7) \qquad \bar{K}^0 = \frac{1}{\sqrt{2}}(\varphi^6 + i\varphi^7)$$

"Tensor" notation

$$\mathcal{M} = \frac{1}{\sqrt{2}}\lambda_\alpha \varphi^\alpha = \begin{pmatrix} \frac{1}{\sqrt{2}}\pi^0 + \frac{1}{\sqrt{6}}\eta & \pi^+ & K^+ \\ \pi^- & -\frac{1}{\sqrt{2}}\pi^0 + \frac{1}{\sqrt{6}}\eta & K^0 \\ K^- & \bar{K}^0 & -\frac{2}{\sqrt{6}}\eta \end{pmatrix}$$

Relation between the currents and fields

Pseudoscalar mesons

$$\partial_\mu j^\alpha_{5\mu} = c_\pi m_\pi^2 \pi^\alpha \qquad (\alpha = 1,2,3)$$

$$\partial_\mu j^\alpha_{5\mu} = c_K m_K^2 K^\alpha \qquad (\alpha = 4,5,6,7)$$

$c_\pi \approx 94$ MeV

Note: Gell-Mann's f_π and Weinberg's F_π are related to our c_π via

$$F_\pi = \frac{1}{f_\pi} = 2c_\pi \quad .$$

The Goldberg- Treiman relation: $c_\pi = -(m_N/G_{\pi NN})(g_A/g_V)$

Vector mesons

$$j_\mu^\alpha = (m_\rho^2/f_\rho)\rho_\mu^\alpha \qquad (\alpha = 1,2,3)$$

$$j_\mu^{em} = (m_\rho^2/f_\rho)\rho_\mu^0 + (1/2f_Y)\left[m_\varphi^2 \cos\theta_Y \varphi_\mu - m_\omega^2 \sin\theta_Y \omega_\mu\right]$$

$$\frac{f_\rho^2}{4\pi} = 2.0 — 2.5, \quad \theta_Y \approx 35^\circ \text{ (current mixing model).}$$

Note: Gell-Mann's γ_ρ (the 1961 Pasadena Convention) is related to our f_ρ (the 1960 Chicago Convention) via

$$\gamma_\rho = f_\rho/2 .$$

KSRF relation:

$$f_\rho^2 = m_\rho^2/2c_\pi^2 .$$

REFERENCES

Chapter 1.

Unitary symmetry in the symmetric Sakata model

M. Ikeda, S.Ogawa and Y.Ohnuki, Progr. Theoret. Phys. 22, 715 (1959).

J. Wess, Nuovo Cimento, 15, 52 (1960)

Y. Yamaguchi, Prog. Theoret. Phys. Suppl.No.11 (1959).

Eightfold way

M. Gell-Mann, CTSL-20 (1961) [reprinted in M.Gell-Mann and Y.Ne'eman, The Eightfold Way (Benjamin, New York,1964)].

Y. Ne'eman, Nucl. Phys. 26, 222 (1961).

Quarks

M.Gell-Mann, Physics Letters 8, 214 (1964).

G.Zweig, CERN 8182 / Th.401 (1964).

SU (2) subgroups

S.Meshkov, C.A.Levinson and H.J.Lipkin, Phys.Rev. Letters 10, 361 (1963).

Chapter 2.

Gell-Mann - Lévy variational method

M.Gell-Mann and M.Lévy, Nuovo Cimento 16, 705 (1960).

Divergence condition

M.Veltman, Phys. Rev.Letters 17, 553 (1966).

Current commutation relations

M.Gell-Mann, Phys.Rev. 125, 1067 (1962)

M.Gell-Mann, Physics 1, 63 (1964)

R.P.Feynman, M.Gell-Mann and G.Zweig,

Phys. Rev. Letters 13, 678 (1964).

Schwinger term

T.Goto and T.Imamura, Progr. Theoret.Phys.14, 396 (1955).

J.Schwinger, Phys.Rev.Letters 3, 296 (1959)

S.Okubo, Nuovo Cimento 44A, 1015 (1966)

Chiral symmetry

M.Gell-Mann, Physics (1964) loc.cit.

Nonrenormalization of vector currents

R.P.Feynman and M.Gell-Mann, Phys.Rev.109,193 (1958).

S.Fubini and G.Furlan, Physics 1, 229 (1965).

M.Ademollo and Gatto, Phys.Rev.Letters 13,264 (1964).

Chapter 3.

Vector mesons and the nucleon structure

Y.Nambu, Phys. Rev. 106, 1366 (1957).

W.R.Frazer and J.Fulco, Phys.Rev.Letters 2,365 (1959).

M.Gell-Mann and F.Zachariasen, Phys.Rev.124,953 (1961).

T.Massam and A.Zichichi, Nuovo Cimento 43, 1137 (1966).

Universality and gauge principle

C.N.Yang and R.L.Mills, Phys. Rev.96, 191 (1954)

J.J.Sakurai, Ann.of Phys. 11, 1 (1960)

A.Salam and J.C.Ward, Nuovo Cimento 20, 419 (1961)

Universality and vector meson dominance

M.Gell-Mann and F.Zachariasen, (1961) loc.cit.

M.Gell-Mann, Phys.Rev.125, 1067 (1962)

Current field identity

N.M.Kroll, T.D.Lee and B.Zumino, Phys.Rev.157, 1376 (1967).

M.Gell-Mann and F.Zachariasen (1961) loc.cit.

Determination of the ρ meson coupling constant

J.J.Sakurai, Phys.Rev.Letters 17, 1021 (1966).

ρ exchange in pion-nucleon scattering

J.J.Sakurai (1960) loc.cit.

J.Bowcock, W.M.Cottingham and D.Lurié,
Nuovo Cimento 16, 918 (1960).

J.Hamilton, High Energy Physics (ed.E.H.S.Burhop,
Academic Press, New York, 1967) Vol.1, p.193.

ρ exchange in nucleon-nucleon scattering

R.A.Bryan and B.L.Scott, Phys.Rev.135, B434 (1964),

A.Scotti and D.Y.Wong, Phys.Rev.147, 1071 (1966).

ρ dominance in ω decay

M.Gell-Mann, D.Sharp and W.Wagner, Phys.Rev.Letters 8,
261 (1962).

Lepton pair decay of vector mesons

Y.Nambu and J.J.Sakurai, Phys.Rev.Letters 8,79 (1962).

M.Gell-Mann, D.Sharp and W.Wagner (1962) loc.cit.

Photoproduction of ρ^{0}

M.Ross and L.Stodolsky, Phys.Rev.149,1172 (1966).

P.G.O.Freund, Nuovo Cimento 44A, 411 (1966),

H.Joos, Physics Letters 24B, 103 (1967).

S.D.Drell and J.Trefil, Phys.Rev.Letters 16, 552, 832 (E) (1966).

Gauge invariance and ρ dominance

G.Feldman and P.T.Matthews, Phys.Rev.132,823 (1963).

N.Kroll, T.D.Lee and B.Zumino (1967) loc.cit.

Gauge field algebra

T.D.Lee, S.Weinberg and B.Zumino, Phys.Rev.Letters 18, 1029 (1967).

Asymptotic behavior of $\sigma(e^{+}e^{-}\rightarrow$ hadrons).

J.D.Bjorken, Phys.Rev.148, 1467 (1967).

J.Dooher, Phys.Rev.Letters 19, 600 (1967).

ω-φ mixing

J.J.Sakurai, Phys.Rev.132, 434 (1963).

S.Okubo, Physics Letters 5, 165 (1963).

S.Coleman and H.J.Schnitzer, Phys.Rev.134,B 863 (1964).

N.Kroll, T.D.Lee and B.Zumino (1967) loc.cit.

Lepton pair decays of ω and φ

R.F.Dashen and D.H.Sharp, Phys.Rev.133, B 1585 (1964).

(See also references for Chapter 6).

Chapter 4.

Troubles with CAC

J.C.Taylor, Phys.Rev.110, 1216 (1958)

M.L.Goldberger and S.B.Treiman, Phys.Rev.110,1478 (1958).

Goldberger-Treiman relation

M.L.Goldberger and S.B.Treiman, Phys.Rev.110,1178 (1958).

Y.Nambu, Phys.Rev.Letters 4, 380 (1960).

M.Gell-Mann and M.Lévy, Nuovo Cimento 16,705 (1960).

σ model

J.Schwinger, Ann.of Phys. 2, 407 (1957)

M.Gell-Mann and M.Lévy (1960) loc.cit.

PCAC and pion pole dominance

Y.Nambu (1960) loc. cit.

J.Bernstein, S.Fubini, M.Gell-Mann and W.Thirring, Nuovo Cimento 17, 757 (1960).

M.Gell-Mann, Phys.Rev.125, 1067 (1962)

Cabibbo theory

N.Cabibbo, Phys.Rev.Letters 10, 531 (1963).

Chapter 5.

Reduction technique

H.Lehmann, K.Symanzik and W.Zimmermann, Nuovo Cimento 1, 1425 (1955).

F.E.Low, Phys. Rev. 97, 1392 (1955).

Chirality conservation and soft pion emission

Y.Nambu and D.Lurie', Phys.Rev.125, 1429 (1962).

Y.Nambu and E.Shrauner, Phys.Rev.128, 862 (1962).

Adler consistency condition

S.L.Adler, Phys.Rev.137, B1022 (1965); Phys.Rev. 139, B 1638 (1965).

Master formula

Y.Nambu and E.Shrauner (1962) loc. cit.

C.G.Callan and S.B.Treiman, Phys.Rev.Letters 16, 153 (1966).

Leptonic K decay

C.G.Callan and S.B.Treiman (1966) loc. cit.

V.S.Mathur, S.Okubo and L.K.Pandit, Phys.Rev.Letters 16, 371 (1966).

S.Weinberg, Phys.Rev.Letters 17, 336 (1966).

Photoproduction and electroproduction

Y.Nambu and E.Shrauner (1962) loc. cit.

S.L.Adler and F.Gilman, Phys.Rev.152, 1460 (1966).

S.Fubini, G.Furlan and C.Rossetti, Nuovo Cimento 43, 161 (1966).

G.Furlan, R.Jengo and E.Remiddi, Nuovo Cimento 44, 427 (1966)

Riazuddin and B.W.Lee, Phys. Rev.146, 1202 (1966).

Divergence condition approach

M.Veltman, Phys. Rev.Letters 17, 553 (1966)

S.L.Adler (1965) loc.cit.

$K_{\pi 3}$ decay

C.G.Callan and S.B.Treiman (1966) loc.cit.

Y.Hara and Y.Nambu, Phys.Rev.Letters 16, 875 (1966)

D.K.Elias and J.C.Taylor, Nuovo Cimento 44A, 518 (1966).

B.M.K.Nefkens, Physics Letters 22, 94 (1966).

S wave hyperon decay

B.W.Lee, Phys.Rev.Letters 12, 83 (1964)

H.Sugawara, Progr.Theoret.Phys.31, 213 (1964)

B.W.Lee and A.R.Swift, Phys.Rev.136,B 229

H.Sugawara, Phys.Rev.Letters 15, 870, 977 (E) (1965).

M.Suzuki, Phys.Rev.Letters 15, 986 (1965).

Y. Hara, Y.Nambu and J.Schechter, Phys.Rev.Letters 16, 380 (1966).

Ward-Takahashi identity

Y.Takahashi, Nuovo Cimento 6, 371 (1957).

Double pion emission

K.Kawarabayashi and M.Suzuki, Phys.Rev.Letters 16, 255 (1966).

Soft pion scattering

S.Weinberg, Phys.Rev.Letters 17, 616 (1966).

Y.Tomozawa, Nuovo Cimento 46 A, 707 (1966)

Adler-Weisberger relation

S.L.Adler, Phys.Rev.Letters 14, 1051 (1965).

W.I.Weisberger, Phys.Rev.Letters 14, 1047 (1965)

W.I.Weisberger, Phys.Rev.143, 1302 (1966).

D.Amati, C.Bouchiat and J.Nuyts, Phys.Letters 19, 59 (1965).

C.A.Levinson and I.J.Muznich, Phys.Rev.Letters 15, 715 (1965).

Connection between current algebra and vector meson dominance

J.J.Sakurai, Proceedings of the Fifth Annual Eastern Theoretical Physics Conference (Benjamin, New York, 1967) p.81.

KSRF relation

K.Kawarabayashi and M.Suzuki, Phys.Rev.Letters 16, 255 (1966) Riazuddin and Fayyazuddin, Phys.Rev.147, 1071 (1966).

Chapter 6

Spectral function sum rules

S. Weinberg, Phys. Rev.Letters 18, 507 (1967).

T.Das, V.S.Mathur and S.Okubo, Phys. Rev. Letters 18, 761 (1967).

Electromagnetic mass difference of pions

T.Das, G.S.Guralnik, V.S.Mathur, F.E.Low and J.E.Young, Phys. Rev. Letters 18, 759 (1967).

ω-φ mixing and spectral function sum rules

T.Das, V.S.Mathur and S.Okubo, Phys. Rev. Letters 19, 470 (1967).

R.J.Oakes and J.J.Sakurai, Phys. Rev.Letters 19, 1266 (1967).